Advance Praise for
MIGHTY BAD LAND

"Explorers who enter the unknown not only learn about a world no other has ever seen before but they also discover themselves."

—**Dr. Robert D. Ballard**, Deep-Sea Explorer, and author of *Discovery of the Titanic (Exploring the Greatest of All Lost Ships)*, and *Into the Deep*, who finally discovered someone he thought he has known for the last fifty years

"Bruce Luyendyk is a highly respected geologist, whose research took him on three expeditions to Antarctica's Marie Byrd Land during the 1990s. He even has an Antarctic Mountain named after him. *Mighty Bad Land* tells the story of the first of these trips to one of the remotest places on earth in personal terms. You emerge with a true sense of the complex emotions and changing relationships that lay at the core of the field experiences. This is a vivid account of the challenges of Antarctic research which resulted in the discovery of a vanished continent: Zealandia. Luyendyk makes tough fieldwork come alive in this unique story."

—**Brian Fagan**, Emeritus Professor of Anthropology, University of California, Santa Barbara, and author of *The Little Ice Age*

"With this intriguing narrative, Bruce Luyendyk has created a new genre, the Geothriller. Humans and their agendas are upstaged by rocks, glaciers, and continents, where all seek mischievous outcomes. Once you read this book, you'll never look at a map the same way."

—**Shelly Lowenkopf,** Emeritus Instructor, University of Southern California, and author of *Struts and Frets*

"In this deeply personal account of his first Antarctic expedition, Bruce Luyendyk weaves a tale of adventure, peril, frustration, and awe. *Mighty Bad Land* is also a fascinating window into the interpersonal dynamics of a small, remote, field party and the complex mind of its author."

—**Edmund Stump**, Professor Emeritus of Geology at Arizona State University and author of *Otherworldly Antarctica*

MIGHTY
BAD
LAND

Mount Luyendyk, Antarctica. (Photo: C Siddoway)

MIGHTY BAD LAND

A Perilous Expedition to Antarctica
Reveals Clues to an Eighth Continent

BRUCE LUYENDYK

Geologist and Namesake of Antarctica's Mount Luyendyk

PERMUTED
PRESS

A PERMUTED PRESS BOOK

Mighty Bad Land:
A Perilous Expedition to Antarctica Reveals Clues to an Eighth Continent
© 2023 by Bruce Luyendyk
All Rights Reserved

ISBN: 978-1-63758-843-7
ISBN (eBook): 978-1-63758-844-4

Cover art by Cody Corcoran
Cover photo: View of Birchall Peaks from Swarm Peak (Chapter 19; Photo: Steve Tucker)
Interior design and composition by Greg Johnson, Textbook Perfect

PERMUTED PRESS

Permuted Press, LLC
New York • Nashville
permutedpress.com

Published in the United States of America
1 2 3 4 5 6 7 8 9 10

To my dear, supportive wife, Susan,
and my clever and brave son, Loren

Stephen F. Tucker (Tucker), 1950–2022.

Always go a little further into the water than you feel you're capable of being in. Go a little bit out of your depth. And when you don't feel that your feet are quite touching the bottom, you're just about in the right place to do something exciting.

—**DAVID BOWIE**, Musician, Artist

CONTENTS

LIST OF MAPS xiii
FOREWORD by Edward J. Larson xiv
PREFACE xix

CHAPTER 1: Point of No Return 1
CHAPTER 2: NPQ 13
CHAPTER 3: McMurdo Station 18
CHAPTER 4: The Lions of Antarctica 27
CHAPTER 5: Radio Room 37
CHAPTER 6: Cape Evans 45
CHAPTER 7: Shackleton's Room 53
CHAPTER 8: Debrief 62
CHAPTER 9: Survival School 67
CHAPTER 10: Recce 78
CHAPTER 11: Put-In 88
CHAPTER 12: Depot Camp 99
CHAPTER 13: South to the Chester Mountains 107
CHAPTER 14: Neptune Nunataks 117
CHAPTER 15: Marian, Ruth, Judy, and Punch 127
CHAPTER 16: Pursuit of Happiness 136
CHAPTER 17: Perfume on Ice 142
CHAPTER 18: Northwest to Birchall Peaks 148

CHAPTER 19: Swarm Peak 154

CHAPTER 20: You've Got Mail 161

CHAPTER 21: Shadows in the Storm................................. 168

CHAPTER 22: Down the Ochs Glacier................................177

CHAPTER 23: Avers Camp 185

CHAPTER 24: Bird Bluff 194

CHAPTER 25: Black Flags 203

CHAPTER 26: Up the Ochs Glacier 212

CHAPTER 27: Pull-Out 220

CHAPTER 28: Out-Brief... 232

CHAPTER 29: The World 239

CHAPTER 30: Finding Zealandia 257

EPILOGUE.. 263

APPENDIX 267
SCIENTIFIC FINDINGS... 275
ACKNOWLEDGMENTS 286
INDEX 288

LIST OF MAPS

1. *Ross Embayment sector of Antarctica showing Ross Island, McMurdo Station, Ross Ice Shelf, and Marie Byrd Land.*

2. *Southwest Pacific showing Antarctica, Australia, South America, and New Zealand.*

3. *Ross Island in the Ross Sea, Antarctica.*

4. *Mountains of the Ford Ranges, mentioned in the story.*

5. *Close-up map showing trails blazed and taken during the 1989–1990 expedition, mentioned in the story.*

6. *Zealandia today.*

FOREWORD

by Edward J. Larson

Over 120 years ago, in 1901, sailing aboard the purpose-built Royal Research Ship *Discovery*, the first major scientific expedition launched to winter-over below the Antarctic Circle headed from England to New Zealand and then over four thousand kilometers further south to Antarctica's Ross Island near the present site of the United States National Science Foundation's McMurdo Station. Captained by Royal Navy officer Robert Falcon Scott and funded by the British government, the *Discovery* carried teams of researchers charged with specific tasks by leading British scientific associations. Chief among their assignments, the expedition's geologists would look for fossil evidence documenting the former connectiveness of the world's southern continents in the deep past and a record of climate change.

At that time, no one knew if an actual Antarctic continent existed. During the 1800s, naval exploring expeditions and some ambitious whalers had discovered land south of the Antarctic Circle, but these isolated discoveries could have represented parts of an archipelago of volcanic islands rather than sections of a continental landmass. Dredging in the deep South Pacific during the 1880s, however, researchers aboard *HMS Challenger* had found sandstone, limestone, and other types of continental rock in sediment deposited on the ocean floor by Antarctic icebergs. This suggested the presence of a South Polar continent.

Further, and of more pressing scientific significance at the time, biologists puzzled over the distribution of fossils among the southern continents. During the 1800s, researchers found fossil evidence of certain extinct species, such as the *Glossopteris* genus of seed ferns, from an earlier geologic epoch in southern South America, Africa, and India. To account for the seemingly abrupt, widespread distribution of such species on now unconnected continents under an evolutionary view of life, Charles Darwin proposed that those continents once were connected in past times, perhaps through an Antarctic continent. Antievolutionists countered that the far-flung distribution of species constituted evidence of their divine creation. Although Darwin died before the *Discovery* expedition, his close friend and chief supporter among British botanists, Sir Joseph Hooker, threw his full weight behind it as a means to bolster Darwinism by searching for *Glossopteris* and other ancient fossils on the supposed Antarctic continent.

Through arduous, death-defying research over three summers and two winters in Antarctica, researchers on the *Discovery* expedition found an ice-covered continental landmass on the mainland across from their Ross Island base, dry (or snowless) valleys among the Transantarctic Mountains, the first documented evidence of sedimentary rocks and fossils on the continent, and widespread evidence of ancient climate change. Antarctica was once warm enough to sustain life, they concluded. Over the following decade, two more British expeditions—one on an aged, refitted sealer named *Nimrod* captained by Ernest Shackleton and another with Robert Scott aboard a sturdy whaler called *Terra Nova*—built on this research, with Scott's second expedition finally finding the elusive *Glossopteris*. Researchers on these expeditions risked their lives on a regular basis, with the scientist who found the *Glossopteris* fossils, Edward Wilson, dying with Scott and his five-person polar party in 1912. After finding those prized fossils at the Mount Buckley outcropping during their descent from the East Antarctic Ice Sheet, this team had carried them until their death from starvation, exposure, and exhaustion on the Ross Ice Shelf. A search party found their bodies in camp with the fossils and field notes.

Despite radios, airplanes, snowmobiles, better supplies, and superior food, scientific research in the remote reaches of Antarctica remains challenging today. As a historian traveling under a grant from the National Science Foundation's Antarctic Artists and Writers Program, I experienced those challenges firsthand working with dozens of different scientific teams during my season on the ice over a decade ago. It was there that I first met Bruce Luyendyk, then on his fifth expedition, whose geological research built on the century-old findings of the *Discovery*, *Nimrod*, and *Terra Nova* expeditions. As a historian, while I worked with various current research teams in the field, my project involved understanding and interpreting the work of those pioneering scientists on earlier expeditions, leading to my 2011 book with Yale University Press, *An Empire of Ice: Scott, Shackleton, and the Heroic Age of Antarctic Science*. Meeting Luyendyk gave me the chance to talk with a modern geologist doing similar work as the *Discovery*, *Nimrod*, and *Terra Nova* geologists, and it helped me to confirm in my mind why their century-old research remains relevant today.

Bruce Luyendyk's memoir, *Mighty Bad Land: A Perilous Expedition to Antarctica Reveals Clues to an Eighth Continent*, recounts the first of his three geologic expeditions to mountain ranges in Marie Byrd Land, eight hundred miles east of the main American base at McMurdo Station. Marie Byrd Land, or "MBL," is a remote area of West Antarctica that stretches east from the Ross Sea. Its scattered mountain ranges are distributed over an area about the size of California. Because Bruce's team worked beyond helicopter rescue range, they found themselves isolated in the Deep Field. Here, for a month and a half in 1989-1990, his small team of six persons sought evidence for the breakup of the Gondwana supercontinent and of a much larger ice sheet on Antarctica in the not-too-distant past, evidence that bears on the current rise of global sea level.

As a memoir, *Mighty Bad Land* presents both a journey of geological discovery and an adventure story. In both respects, this book is similar to the memoirs written by Scott, Shackleton, and some of their leading scientists about their tales of living and working in the Antarctic. Like the scientists deployed by Scott and Shackleton, Luyendyk's team had to prepare to survive on their own. They were challenged, and sometimes unprepared for, the obstacles they faced: biting cold, screaming blizzards, whiteouts, falls into deep crevasses, and bitter interpersonal conflicts. Like Shackleton's lead field scientist, Royal Society geologist Edgeworth David, Luyendyk was nearly fifty when he began his first Antarctic expedition. Burdened by age and asthma, he (like David) chose to risk the journey.

Luyendyk's team consisted of four geologists supported by two mountaineers. Their combined skills and expertise were many and in some cases lifesaving. The geologic team included a graduate student researching the origin of these mountains for her dissertation. Together, they lived and worked, discovered, and protected each other from lurking perils.

Today, ice-covered Antarctica sits centered over the South Pole at 90 degrees south latitude. New Zealand lies far to the north, halfway to the equator at 45 degrees south latitude. In the deep past, however, Antarctica and New Zealand were part of the same supercontinent, "Gondwana." The early geologists traveling on the *Discovery*, *Nimrod*, and *Terra Nova* expeditions envision connections between these lands but, assuming their locations fixed for all time, supposed that land bridges linked them in earlier geologic eras. Scientists today see continental drift as causing past connections and subsequent divisions. Both then and now, these scientists sought to test their ideas in the field. Even as the scientific answers change, the scientific method for seeking those answers remains much the same. Always questioning, always testing, no matter the risks, science advances one hypothesis at a time. A shared passion for truth—for the best answers—drives these researchers. Their stories make for compelling reading.

New Zealand itself consists of two large islands and some smaller ones that stand as exposed parts of an expansive shallow subsea feature named the Campbell Plateau. Recently, geologists have identified other submerged plateaus nearby that comprise sunken continental pieces of Gondwana. New Zealand represents a small part of the vast submerged regions that Luyendyk named Zealandia in 1995 while analyzing the results of their Antarctic expeditions. In recent years, scientists have confirmed his theory of this sunken continent using the name he gave it.

Antarctica holds significant public awareness. Thirty nations maintain research stations on Antarctica, bringing its summer resident population to some four thousand persons, almost all scientists and their support staff. Only a precious few of them ever reach Marie Byrd Land and most no longer face the dangers confronted by Luyendyk and his team. *Mighty Bad Land* tells the story of this modern team in a manner reminiscent of the old tales of derring-do of early Antarctic scientists who faced similar perils. This makes *Mighty Bad Land* a throwback of sorts.

Unlike most recent books on Antarctica, the reader becomes embedded with Luyendyk's team and experiences firsthand the challenges, companionship, failures, bravery, cowardice, and success that scientific research in an unforgiving place brings to light. Bruce takes readers with him to a pure wilderness experienced by few humans—a place where unseen menace waits everywhere. Readers will learn why Marie Byrd Land, named for the wife of mid-twentieth-century Antarctic aviator Richard Byrd, has earned the nickname Mighty Bad Land.

Mighty Bad Land is an unvarnished account of a scientist and his team exploring one of the most remote wild places left on Earth. Now retired, Luyendyk tells this story firsthand. Readers will learn from him what kind of people do science at the uttermost end of the earth and how they do it. They will follow his team, and see his personal challenges, on their first expedition. They'll find the answer to the question: what does it take to prevail in Antarctica today? It takes the same sort of grit that it took 120 years ago from scientists traveling on the *Discovery*, *Nimrod*, and *Terra Nova*. Welcome aboard.

PREFACE

|||||||||||||||||||||||||||||||||

Ice is the beginning of Antarctica and ice is its end. As one moves from the perimeter to the interior, the proportion of ice relentlessly increases. Ice creates more ice, and ice defines ice.

—**STEVEN PYNE**, *The Ice: A Journey to Antarctica*

The ocean itself is blue, but a thick white ocean of ice buries the continent of Antarctica. Near the continent's margin, mountains jut through the edge of the ice to reveal the secrets of what lies beneath.

What does it take to discover these secrets? What drives a person to take on such a quest? What does Antarctica demand in return? My memoir is an exploration of the inner man in the most remote place in the world, where zero degrees Fahrenheit is a warm day, grandeur stirs disbelief, and in summer, there is no night.

On January 6, 1990, wracked with fear, I stood alone on the snow of Marie Byrd Land in West Antarctica. I watched, helpless—in the distance, members of my small team attempted a rescue of two of us who had disappeared into a crevasse. We didn't intend to risk our lives while we explored the remote reaches of the Ford Ranges. Near misses happened with blizzards, ice falls, and crevasses.

Our team of one woman and five men sought secrets of an intriguing geologic event that formed a sunken continent long hidden beneath the waters of the South Pacific. A continent of scattered pieces, including the nations of New Zealand and New Caledonia, now known as Zealandia. In Marie Byrd Land, that is, Mighty Bad Land, we searched for evidence of how and why large bits of Gondwana ripped off from this part of Antarctica and then submerged as they drifted north. My project role: the expedition leader. We'd secured funding from the US National Science Foundation to solve this puzzle. It had taken two tries over a span of two years to get approval. Now we found ourselves eight hundred miles from the main base at McMurdo Station—alone.

Imagine flying from Los Angeles to Denver, passing over a vacant frozen wilderness, arriving to find nobody and nothing there, then living and working on your own in the wilderness. Remote expeditions such as ours carry the label Deep Field. That means beyond helicopter rescue range. Unlike in Antarctica today, we didn't have GPS, internet, faxes, satellite phones, or even walkie-talkies. We didn't have resupply. Our only link to humanity: a twenty-watt short-wave radio.

After the NSF approved what would be our two expeditions, I discovered more about the scale of effort we'd need. Information trickled in from Antarctic veterans that showed me the complications, risks, and dangers ahead. Their message: in Antarctica, mistakes are not forgiven.

I'm asthmatic. The NSF rejected me for this project until I took a cold-air exercise test to prove I could tolerate the conditions I would face. These conditions included exertion in cold, dry air—known triggers for asthma attacks. After some time in the mountains, my anxiety for our safety threatened to overwhelm me. Afterward, I met with our lead mountaineering guide; he told me not to come back for the next expedition—I didn't have the steel to deal with the danger and chaos of Antarctica. I considered this advice, but in the end, resolved that I was the leader, I faced my fears, and returned.

In time, I realized the splendor of what we saw in the wilderness of ice, rock, and sky, what we lived through and became part of. I fell in love with a continent.

In this story, I deal with scientific topics at a level for a lay reader. I reveal them in the dialogue and portray them with our actions. We made discoveries—about the link between mountain building in Marie

Preface

Byrd Land and the origins of Zealandia, about a larger West Antarctic ice sheet that existed in the past. Today, research in the Antarctic has a sharp focus on climate change. At the time of this story, 1989–1990, plate tectonics occupied the forefront. Climate change had only started to get popular recognition the year before, with the testimony of James Hansen to the US Congress. Nevertheless, we encountered effects of a warmer Antarctica even in 1989.

The events described in this book comprise my recollection and my interpretation. I have used my personal journals, notebooks, and photographs taken by me and by others on my team to build the foundation. In places, I may have mistaken one person's actions for another's. I changed the names of a few people or did not state them.

I usually did not record dialogue in my journals. However, I recall some of what we said to this day. I affirm that most of the dialogue in this book and those I attribute it to fit my memory; in some cases, they could have said it as it was appropriate to them and the situation, but it may not be so, or they may not have said it exactly as I remember it. A few scenes are moved in time and place. This book is not a work of journalism. I represent the actions of myself, my team, and others to the best of my documentation and memory.

Unless otherwise noted, all photographs are by the author.

In the back of the book, I have sections explaining some issues of Antarctic geography and orientation: rocks, minerals, and formations mentioned in the story; and acronyms that appear throughout. A section entitled Scientific Findings presents our discoveries at a more scientific level than covered in this story, with a bibliography of scientific publications where these findings appear.

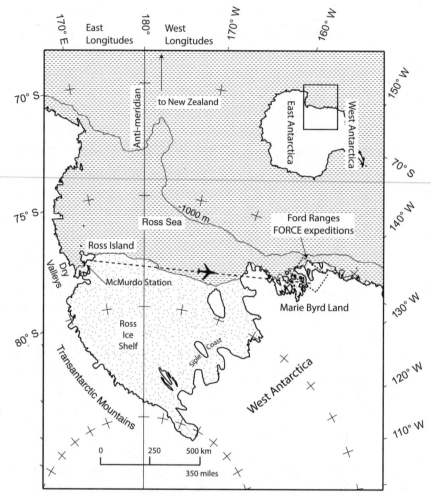

MAP 1. *Ross Embayment sector of Antarctica showing Ross Island, McMurdo Station, Ross Ice Shelf, and Marie Byrd Land. The dotted box outlines the region in the Ford Ranges of Marie Byrd Land where the story takes place. The dashed line shows air route to base camp in the Ford Ranges, eight hundred miles from McMurdo. This map shows the antimeridian (180° longitude) on the opposite side of the globe from the prime meridian at Greenwich, England, and the distinction between east and west longitudes, which is why Marie Byrd Land is in West Antarctica. (After US Geological Survey publication)*

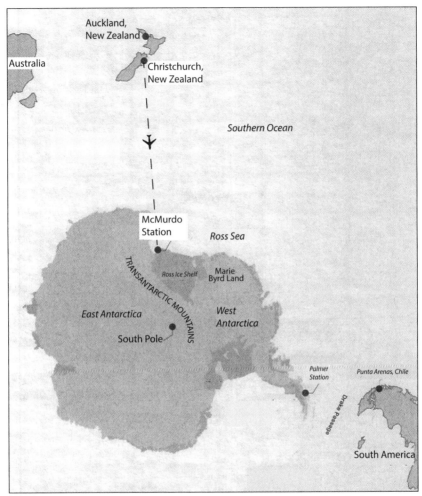

MAP 2. *Southwest Pacific showing Antarctica, Australia, South America, and New Zealand. The US Antarctic Program participants acquire cold weather gear in Christchurch, then fly over 2,400 miles south to McMurdo Station, the main US Antarctic base located on Ross Island at the edge of the Ross Ice Shelf. This story takes place in the mountains of Marie Byrd Land in West Antarctica, about one thousand miles east of McMurdo. (After US National Science Foundation publication)*

CHAPTER 1

Point of No Return

We have two or three great and moving experiences in our lives—
experiences so great and moving that it doesn't seem at the time
anyone else has been so caught up and so pounded and dazzled and
astonished and beaten and broken and rescued and illuminated
and rewarded and humbled in just that way ever before.

—F. SCOTT FITZGERALD

Christchurch, New Zealand, to McMurdo Station, November 21, 1989

Places exist on Earth where no human has gone. Many are unknown, or invisible and inaccessible, like the floor of the ocean. Places of unspeakable beauty, frightening to our imagination. They draw us to them, as if by an extra pull of gravity. They mesmerize. Few of us get the privilege to visit; the opportunity to explore such places rarely presents itself. To go to them is hard work. At these places, sudden menace can confront and test you. These places change your life forever—you never really leave them. The last place on Earth is Antarctica. I had nurtured a dream of going, decided to take my chances, and now I was on my way, charging south, flying on a Hercules to the bottom of the world and The Ice.

1

"I want to see Antarctica from the cockpit," I shouted to Tucker, Steve Tucker, my friend and one of our mountaineers, who sat next to me. "We must be close now." My message to Tucker had to penetrate through the yellow foam earplugs we wore to suppress the din. I timed my visit to the flight deck so I could see Antarctica for the first time. I put down *Texas* and twisted out of my seat to stand up and stretch out the kinks.

We and twenty or so passengers were planted in red web seats on fold-up aluminum frames that stretched along both sides of the Hercules cargo hold. We sat on these, crammed hip to hip and knee to knee. A parallel center row in the aircraft held a dozen more, including Steve and Dave, two of our expedition team, across from us.

Dave Kimbrough sat in the center row asleep, his large knees arranged to avoid mine; his red parka pulled over his head made a small tent. The guy could sleep anywhere. Lucky him. Tucker and Dave seemed relaxed, reading and napping. I wasn't. As the leader of my first of what would be two expeditions, my mind swarmed with several to-do lists, most of which proved later to be off target.

Dave and I, geologists and professors, planned this research project a few years back. It took us two years to get approval and funds from the US National Science Foundation (NSF), a federal agency, to get to this point. I reflected on who had committed to the project, to me, to our team: the NSF, Antarctic program staff in the US and Antarctica, US Navy aircraft and crews, our university, and our families. I didn't attempt to put my mind around the total cost; it was a lot. These facts awakened the burden of the expedition on me, the scale of the investment made by those who decided to take a chance on us. We wouldn't get a do-over, but we'd be held to account—like the NASA aphorism, "Failure is not an option."

Now we began our mission: to explore and discover. To find out when, how, and why the southern supercontinent of Gondwana split apart—leaving Marie Byrd Land (MBL) in Antarctica while scattering continental pieces, including the island nation of New Zealand, across the South Pacific. Important parts of the story demanded answers: Did the Earth's crust stretch so much during the split as to build the mountains in MBL? Why had mountains formed there while the continental pieces on the New Zealand side sunk below sea level? Why was the result not the same? Understanding the plate tectonic processes required answers to these fundamental questions. We, as geologists, planned to unearth the secrets of Marie Byrd Land to find them. We'd do that in the mountains of the Ford Ranges, far into the wilderness of Antarctica.

I stepped between the knees of the other passengers who sat alongside and across from us and turned each foot sideways to slip my oversized white rubber Bunny Boots between legs. With each step, knees bumped knees. I made my way forward.

At the entrance to the flight deck, I met the legs of a crewman above me. "Hello, can I come up?" He waved okay. The Hercules, a New Zealand Air Force C-130 with four colossal turboprop engines, could glide through the sky like the best of them. Our aircraft approached Antarctica from over the Southern Ocean.

I climbed up a few steps. A brilliant sky met me. After the murky cargo hold, my eyes needed time to adjust to the blinding brilliance, my ears to the quiet of the flight deck. The aircraft floated in the sky. Excitement caught in my breath.

The pilots sat at the front of the cockpit, the engineer behind and between them in a jump seat. The navigator sat at an instrument panel to the right rear. He looked up at me and smiled. What a bunch of good-looking guys, young and healthy—made me feel a bit insecure, pushing fifty years old and trying to hide, even ignore, my asthma. They wore green uniforms with the New Zealand Southern Cross flag on the shoulder, headsets for communication, and of course, aviator glasses. Flight instruments covered ceiling and wall space. Floor-to-ceiling windows wrapped around the cockpit, yielding wide, startling views. Immediately, I sensed a moment of inspiration ahead.

The pilot dropped our altitude, then flew the Hercules over the coastline. An inconceivable expanse of white stretched ahead and beneath us. Ice reached beyond the horizon—pure snow below a vivid blue sky. Glaciers spread out like rivers and tributaries; steep, pointed, dark, majestic peaks jutted up from the blanket of snow and ice. I imagined myself as the first person to discover this amazing scene. I became still, lost in that thought.

In the distance, the ice dominated. The immense Antarctic Ice Sheet buried the highest peaks, ignored their existence, made them disappear. I visualized the continent without it. Ice wasn't always here; it started to grow and then bury the landscape over thirty million years ago. The continent had sucked up a good part of Earth's oceans to form this white blanket. Sea level had dropped over two hundred feet to build the ice. What triggered that change, from a warm greenhouse Earth to the present-day icehouse Earth we live on? I dwelled on that deep mystery.

Soon I would step foot on the landscape below, as a geologist and explorer, finding answers to fundamental questions about the breakup of continents. A thrill came over me—mixed with a wisp of dread. I flashed for a moment on the stories I had read and heard from the veterans, the warnings of danger and death.

The scene before me looked alien, not of Earth. So overpowering was this vast, vacant expanse, a white ocean of ice with scattered islands of rock. This moment so special, this sight a beauty beyond explanation, its impact on me profound and not expected. I knew my Antarctic adventure would not be topped in my life. I knew I'd carry it with me forever. A warmth flowed into me with that realization.

I recalled my elementary school teacher's finger scanning various places on a world map hung in front of the chalkboard: continents, countries, capitals. I knew them all at the time. I noticed a thin, ragged white strip running along the bottom of the map, the coast of Antarctica, and wondered what more lay beyond the bottom edge. At recess, I inspected the map up close.

"Where's the rest of the world down here?" I asked. She pointed to a small circular inset map in the lower corner.

"If you could fly above the bottom of the world and look down, you would see Antarctica, a land covered in ice."

"Wow. How big is it?"

"As big as the US and Mexico put together." She pointed to those countries.

Antarctica looked like an enormous white duck. The Antarctic Peninsula jutted out, forming a duck's bill to nibble on the tail of South America, the duck's back the south edge of the Pacific Ocean, its belly facing the Indian Ocean. Many parts had the label *Unexplored*. *Marie Byrd Land* labeled the duck's head. "Who's she?" I asked, pointing to the name.

"That's Admiral Byrd's wife," she said. "He discovered that part of Antarctica in 1929. He named it after her."

Reflecting on this memory, I thought about what drove our team now. Discovery. Although satellites and aircraft had largely imaged Antarctica, many regions lay untrodden, pristine. The same excitement that led me to explore the ocean floor earlier in my career pulled me to Antarctica. New knowledge waited for us at every step. Very few places on Earth offer an experience like scientific exploration in Antarctica.

Others waited to see the view from the flight deck, so I stepped down and threaded my way back to my seat in the dark and noisy cargo hold. The cavernous interior had a subterranean feel to it: dim, and jammed with scientists, staffers, and aircrew, plus cargo. Mostly men, a few women, filled the web seats. I guessed their ages to be from twenties to a few sixties. Crewmen in the hold wore olive drab parkas and jump suits. The passengers dressed alike in the same anonymous clothing— red parkas, plaid wool shirts, black wind pants, and goofy white Bunny Boots. When boarding the plane, I had seen that Tucker exchanged his Bunnies for lightweight blue canvas FDX boots. Why hadn't he given me a heads-up? Every man for himself? Too late for me to swap. All the passengers aboard headed south for various science projects, perhaps oblivious to the dangers below and the adventure ahead. I tried to read, but a seesaw of emotion between excitement and anxiety kept me distracted.

The Spartan interior had a tailgate ramp at the rear that spanned the entire width of the plane. Inside, everything lay exposed—ribs of the airframe, wires, hoses, and pipes. The Herc had a few portholes along the sides, behind the red web seats. People, or clothing and baggage, blocked most of the view. Each side of the hold toward the rear included doors for paratroopers.

People slept, or read, or ate bagged lunches given to us at departure. Conversation of more than a few words couldn't happen—too loud. Strangers leaned their heads on neighbors' shoulders to nap. Others squirmed on occasion, sought unavailable comfort. Many passengers had flung off their parkas, not that cold now, and stuffed them in corners and spaces behind the red webbing. The crew had stacked cargo high toward the back of the plane. The passengers sat forward. Some folks had made their way to the rear cargo area, stretched out over boxes, appeared to nap.

Four of our party—Steve Richard, Dave, Tucker, and I—rode this eight-hour flight from Christchurch to McMurdo Station, Antarctica. Over the past few days, we had learned about the OAEs, or Old Antarctic Explorers, who had visited The Ice many times, and about the FNGs, or "fingees"—Fucking New Guys. Scientists had their own name—Beakers. My own team of six, counting me, included both FNGs and Beakers.

I was the science principal investigator, the PI—a leader of equals, except I was the oldest by far. I had sent my head mountaineer, Alasdair Cain, and my new graduate student, Christine Smith—we called her Chris—to McMurdo about ten days earlier. Now in McMurdo, they organized our gear and supplies, our loads for the Put-In flight to the Ford Ranges in December. Well—that's what I hoped they did. They didn't seem to need or take direction from me so far. Always a step ahead of me, Chris worked like a blonde tornado. This expedition was her graduate project. She was all in. I wanted to trust her. If I didn't, there would be conflict.

Cain, a Scotsman, and a licensed Alpine guide, had thin pale brown hair and a wiry build. I predicted he was brave. People who don't brag often are. Tucker and Cain had not visited MBL before; none of us had. Only Cain had participated in Antarctic expeditions. Cain seemed to tolerate this upcoming experience with amateurs. I wasn't sure of his regard for Tucker; he wasn't free with his opinions.

Steve Tucker, my friend from Santa Barbara, had taught me some rope techniques on the cliffs of the backcountry; I did well at it. Besides that, he had climbed El Capitan in Yosemite, a feat I couldn't imagine. Tucker loved the idea of our upcoming journey to the empty wilderness, showed it in his steady cheerful manner. I hoped that Cain and Tucker could work together.

My skill set included geomagnetism. I'd use it to determine the motions of the continental pieces scattered across the South Pacific in the breakup of Gondwana long ago. Dave would do geochronology—determine the ages of the rocks we found to set the timing of events. Chris and Steve Richard formed a strong team; technical experts and a platonic duo, they understood metamorphic rocks, those formed under high heat and pressure from other rock deep in the Earth's crust. Both could display a passion for arcane topics of metamorphism. They would figure out how far the mountains in MBL had risen from the depths and when. Their technical conversations lost me after a few words.

Steve, next to Dave, shifted in his seat, pulled at his thin black beard but didn't stop reading a geology paper; the guy was always on the job. Steve didn't seem to realize he would soon land in Antarctica, his mind so focused on our geologic goals. The black wind pants with six pockets looked baggy on him, tight on Dave. Steve's dog tags lay across his chest on his black-and-red wool plaid shirt, like all of us wore. I recalled when we had gotten these tags at the US Antarctic Program Clothing Distribution Center, a dilapidated warehouse at the Christchurch airport—the CDC of the USAP. I'd had to learn a bucket of acronyms.

* * *

At the CDC in New Zealand, on a warm November summer day, we had acquired Emergency Cold Weather clothing—ECW. Everyone on board wore it; the aircrew had their own version. Two orange bags, each the size of two pillows, held ECW items, jammed to fit inside. I suspected that they included a lot of military surplus items; they had that kind of musty smell and oily feel. OAEs had forewarned us. We planned to replace almost all the issued items with high-tech mountaineering

Bruce boards Hercules in Christchurch, New Zealand. (Photo: Steve Tucker)

clothing, plastic boots with slip-in liners, and climbing hardware we'd purchased before departing the US. We kept the unique red goose-down parkas, aka the Big Red, the lifesaver: a sturdy, red-hooded parka with enough pockets for Captain Kangaroo. I slipped it on—a modest struggle—and admired its robust construction. It fit well, what I needed to protect myself. I inspected the item, searched the pockets inside and out. The parka had twelve pockets—enough to exasperate me later when I searched for crucial items during a blizzard. *Where did I put my heavy gloves, my goggles, my camera?*

At the bottom of one of my orange bags, I found the dreaded Bunny Boots. I cringed, then yanked them out. White, rubber, and inflatable for insulation, they looked like dead ringers for Mickey Mouse's wardrobe. They made me feel like a little kid again, or a clown about to take a pie in the face. I tried them on and stumbled a few steps, still in my boxers. "These look totally stupid and I can't walk in them," I said to Tucker. He shrugged. *Did anything bother Tucker?*

"Better than lost toes, huh? Heard they're rated to minus sixty-five," he said. I decided to believe him. I knew toes could freeze and die with frostbite, then gangrene took hold. I accepted my new footwear.

The Center staff handed each of us a large manila envelope containing instructions on when and how to report for the eventual flight south and other miscellaneous information. I reached inside this envelope, labeled with our project number, S-070, or Sierra Zero-Seven-Zero. That's what our team would be known as in the USAP. Startled, I found two dog tags. Steve and I looked at each other as we pulled them out. I had never been a soldier, but I knew their purpose.

"Hey, Steve, dog tags in case they have to identify our dead frozen asses."

He dug his out and dangled them. They jingled. "Oh, yeah." He gave me a crazy grin, put the chain around his neck like an award, smiled, and looked down to admire them. They flashed in the low light of the warehouse. One tag hung on a ball chain long enough to go around my neck. It would hang there for more than two months. But, attached to the long chain, a short four-inch chain held the other tag. In the event of my death, a Search and Rescue team would collect this short chain from my corpse.

Chris and Cain got equipped at the CDC the week before. The original plan called for the rest of us to follow them in a few days, but poor weather in McMurdo caused the cancellation of our flight four times over a week. When the weather cleared at last, we trekked to the airport and suited up in ECW at the CDC. Our weights with gear were taken, and a dog sniffed us for drugs. Then we loaded onto a bus to head to the Hercules we'd fly in. After a few minutes waiting on the bus on the hot tarmac at Christchurch, we heard about the Point of No Return.

We learned about the PNR as we sat sweating in our ECW inside that bus. Our young, athletic Kiwi Air Force pilot entered the front of the bus to brief us about the flight. Part of his speech included the Point of No Return. In recent years, the US Antarctic Program changed the name to Point of Safe Return over objections that the label PNR sounded too melodramatic. I had heard rumors and talk of the PNR.

I didn't like the concept. The flight time would take a bit more than eight hours, most of it over the Southern Ocean. "If we have to ditch, follow orders of the crew; they'll help you get into immersion suits and life rafts," he said. "Otherwise, you'll last about five minutes in the water." I'd worn these suits before on research ships—made you look like a red Gumby. *Why bother to ditch? If we're gonna crash, why not nosedive into the sea and get it over with? We can't be rescued anyway. No ships around.*

"The PNR is about four hours out. We'll call McMurdo and check the weather, see if we can land on the Sea Ice Runway. Can't land at Williams Field; our aircraft doesn't have skis," the pilot said. "Weather's okay, we continue; not good, then we boomerang back to New Zealand." Huh, a trip to nowhere and back. "Past the PNR, we can't return—don't carry enough fuel."

Someone on the bus asked if the runway had instrument landing aids. "No, we use Visual Flight Rules," he said.

A man next to me with a sweaty face muttered, "Could need to land in a whiteout; that's happened more than once." I wondered if this pilot had done one of those.

"Do you know about one?" I tested his knowledge. I cleared my throat twice.

"*Pegasus*. Ask around," he said.

He paused, then decided to tell me.

"A Super Connie named *Pegasus* passed the PNR in the early seventies." He looked at me. "Then a whiteout and blizzard came in at McMurdo. But they had to keep goin' south, not enough fuel to go north. They made a bunch of passes tryin' to land, couldn't see a thing. Ended up clipping a wing off but got down."

"Anybody killed?" My stomach cramped.

"No. Crazy huh? Wreck's still there, mostly buried. You can get to it by snowmobile. Tails stick high out of the snow. One or two of them left now, easy to find," he told me, then smiled. I nodded. I didn't want to visit that.

* * *

A sleepy Bruce inside Herc on flight south. (Photo: Steve Tucker)

Now about six hours out of Christchurch, the PNR behind us, we had two more to go before landing. We had to land regardless of the weather in McMurdo. I wondered about the forecast. During my visit to the flight deck, I had seen a low cloud layer over what I guessed was McMurdo Sound, our destination. The aircrew had not mentioned any concern.

Our flight continued south beyond the coast. The pilot made a small course change to get on approach to McMurdo. People in the hold got restless, struggled to move, stretched, dug into their packs, arranged their stuff.

The interior of the plane darkened; the sky had become overcast. The loss of sunlight seemed ominous. I squirmed in my seat trying to peer out a porthole near me, but I only saw gray. The pilot changed course. I noticed that the faint orbs cast on the walls from the portholes moved as we turned. I tried to guess what course we followed. I thought we dropped in altitude, but I didn't know for sure until my ears cleared. Inside the Herc, passengers pulled on their parkas, hats, and gloves, buckled up, prepared for our approach and landing. *Is there a whiteout?*

Are we safe to land? What will it feel like to land on floating ice? I'd have answers soon.

Course and speed changes came abruptly, causing some people to look up and around. The power on the engines decreased, and I recognized the sound of the wing flaps extending. A faded, dirty fasten-seat-belts sign lit up on the front bulkhead. A crewman walked between our knees, checking that we had buckled. He yanked on a few belts to double-check.

The interior now darker, I heard the landing gear whine as it cranked down. The Herc glided with a slight backwards tilt. I decided that the ice landing strip must be in sight. After a peaceful minute, a firm bump told me we had landed—with grace, on the sea ice. I realized I had held my breath and let it out. The aircraft came to a sudden stop that pushed us all sideways and forward in our seats and launched a few loose items.

Crewmen wearing headsets on long cords moved toward the forward door on the left side near my seat. The pilot shut down all engines but one. The quiet exposed the high-pitched whine of hydraulic motors inside the plane. *Hey, I'm in Antarctica.*

In a moment, the door opened. The crewman at the door looked out, exhaled; his breath made a white fog. He pulled up his parka hood. In the rear of the Herc, the crew let down the tailgate ramp to bring in bright, hazy light and prepared to unload the cargo. An immediate bitter cold invaded the plane. No one spoke. Many acted like they knew what to do next. The rest of us took the lead from the OAEs. We stood up, turned, and marched forward single file. People in front of me disappeared out the door; they dragged their orange bags.

My parka zipped and its hood up, I ducked into the open doorway. Colorless cold struck me. I looked to my feet, to the four steps I had climbed up eight hours earlier in green, warm New Zealand. I stepped down them into Antarctica.

CHAPTER 2

NPQ

I swear to you that to think too much is a disease,
a real, actual disease.

—**FYODOR DOSTOYEVSKY**, *Notes from Underground*

Santa Barbara, August 1989

"A bit faster, Professor," one of the two doctors said. I quickened my pace on the exercise bike. A clip pinched my nose shut. I breathed air at zero degrees Celsius through a mouth tube—the cold-air challenge. I wore only shorts and tennis shoes. Doctors had glued wires with electrodes to various shaved places on my chest. My sort-of new girlfriend wouldn't like those bare spots.

"Almost there; need eighty-five percent of your maximum predicted heart rate. Push it up to one-fifty," the second white-coat doctor said. He watched my pulse on a monitor. The first doctor looked at a chart as it printed out. One-fifty—I'd do it, even if I could pass out. I tightened my grip. I went for it.

* * *

A week earlier, news had come that I'd been rated NPQ, or Not Physically Qualified, for the US Antarctic Program. In late spring, my head full of foreboding, I'd filled out a medical questionnaire with a list of sixty conditions. That form came with a packet of instructions for required medical and dental exams along with a Participant Guidebook that included advice to make a will. *A will—really?*

"Do you have or have you ever had any of these: asthma, yes or no?" Yes. That question had also appeared on the NASA astronaut application I filled out a few years before. I'd been disqualified at the start for it.

My asthma seemed well controlled to me. I'd begun to take it seriously a few years back. That started with a phone call.

"Bruce, this is Linda. Are you okay?" my ex-wife announced. I had returned from visiting our son, Loren, at her house. "I saw you using your inhaler and gasping most of the time you were here."

If my ex is concerned, I'd better pay attention.

"You might want to see a doctor very soon," she advised. I took her advice.

"Bruce, you're at seventeen percent breathing capacity," my doctor said. "I'm putting you on steroids to get you out of the woods and prescribing some medications you need to take every day from now on, without fail."

"You mean I'm in trouble?" I asked.

"Yes. Very much so."

The meds worked. A lightness filled me. Since then, I hadn't been to the ER. In the past, I'd gone once or twice a year.

I appealed my NPQ decision. USAP told me to take the cold-air exercise test—see if I could handle maximum exercising while sucking in frozen air, like in the Antarctic conditions I would face.

* * *

I pedaled, determined, my thoughts focused on success. "Not much longer," the doctor said. "Okay, you did seven minutes at eighty-five percent. Good job. Slow down now." I got off the bike, sat on a plastic chair, panted, and sweated. They had me blow out as hard as I could into a measurement device every few minutes. An urge to throw up gripped me. *What will I do if I fail this?*

The answers came back in a couple of days. My lung doctor wrote to the USAP doctor: "Results show…ruling out cold-air-induced obstructive airways disease…I have no hesitation recommending him for the Antarctic program…."

I got that checked off—a weight floated away—but I had to jump one more hurdle. I'd checked yes to the "depression or excessive worry" question on the medical form. I figured, who hasn't been there? That also got me NPQ. I had to appeal and repair the damage from my stupid honest answer. I went to see the doctor who had treated me two years earlier for insomnia and anxiety. He showed more interest in my asthma. He looked at my file, the cold-air test, and my medical history.

"Your asthma diagnosis has been moderate to severe airway obstruction more than once," he told me. "That's serious. You could die in Antarctica. You know that, yes?" He looked straight at me and pointed to the comments in my chart.

"I think I'm more under control now," I said, as a leaden sensation filled my stomach. I remembered the ER visits. Then, I recalled what was at stake: a priceless chance to explore a remote hidden land, to discover, to change the plate tectonic story of the bottom of the world. I ignored his warning.

The doctor wrote a letter to USAP explaining my circumstances when I had gone to see him a couple of years back:

At that time [spring 1987], *in the context of an impending divorce from his second wife, his mother's cancer and her sudden blindness in one eye, a drug addict stealing $34,000 from his mother, and losing custody of his son Loren who went to live with his mother…he developed the following symptoms: Difficulty falling asleep, early morning awakening, sadness, blueness, some diminished concentration, mild anergia, anorexia with a 10 lb. weight loss and mild anxiety attacks… Nevertheless, despite his manifold psychological stressors…he continued to carry a full teaching load, do research and publish.*

Back then, my life was a mess. I had done well through those struggles. I gave myself a deep, satisfied breath. Recalling his letter had brought up things I'd buried away.

His father died when he was 10 years old leaving him the master of the house and having to take care of too many things too early. When he died 'It hurt me a lot,' he quoted me.

That hurt again—to read my own words about my father. Suddenly, a wave of coldness hit me. I remembered the horror of his death.

My father stood at the front door of our home, ready to leave for the last time. His right hand on the doorknob held it partway open. He looked up at me where I sat on the top of the stairs outside my bedroom. I wore my pajamas. He had stuffed his nostrils with now blood-stained white cotton. With a half-smile, he raised his left hand up to his shoulder and waved goodbye—off to the hospital. I waved back to him. He'd gone to the hospital last year and come back. This time, a sob stopped in my throat.

I was at the hospital when he died. I sat with my two brothers in the back seat of my uncle's car, a pale green Dodge sedan, in the parking lot outside Dad's hospital room on the ground floor. My uncle and grandfather got out of the car and went in to visit, my mother already inside. My brothers and I waited—not old enough to go in Dad's room. They returned. My uncle was starting the car to drive us home when my mother screamed from the window of my father's room: "Eager, come back; Pete just died!"

My first experience with raw terror took place. My breath became shallow and rapid. Uncle Eager stopped the car. I moved to the edge of the seat. My uncle rushed inside to my father's room. More cries and screams burst out the window. He came back in tears, wiped his eyes with a white handkerchief.

"Your dad died," he said to my brothers and me. He got in the driver's seat.

"What?" my grandfather asked. He adjusted his hearing aid.

"Pete died."

My grandfather stared straight ahead.

I dropped my head, leaned my arms against the back side of the front seat. I let out the sobs that I'd held in for the past few minutes. My two younger brothers sat stunned, not sure what had happened. I knew. No more help with building model airplanes or my electric trains. No more silly cartoons he'd draw at the kitchen table. No more baseball games. No more watching him leave for work—tall, erect, a gray homburg hat his signature.

Later, my mother arrived home alone. I met her on the front steps and hugged her. She said, "You're the man of the family now; you have to take care of your brothers."

"Yes, Mom, yes, I will," I said, then sobbed with her. I was ten, adrift, lost in pain.

I stared at the doctor's letter. What would Dad think of what I'm about to do? Would he feel excited and proud, think I'm foolish, try to stop me? I knew the answer—what a man who came to this country on his own and in his youth would say.

Near the end of his letter, my doctor wrote:

Assessment: History of major depression with mild anxiety attacks, treated and in substantial remission. In the absence of new severe psychological stressors…his prognosis for a complete remission of his depression and anxious symptoms is excellent.

I passed—I got PQ-ed. I congratulated myself. Still, I regretted that this deep personal view of me got shared with the USAP bureaucracy. I didn't have other options. Looking back on these past events that almost brought me down, I realized that I proved both vulnerable and resilient. I knew that last trait of mine would carry me through this adventure. I felt strong and determined.

CHAPTER 3

McMurdo Station

The first time you come down for the adventure. The second time for the money. And the third time because you can't function anywhere else anymore.

—**ANONYMOUS WORKER** at McMurdo Station

McMurdo Station, November 21, 1989

My Bunny Boots almost slid out from under me as I attempted to walk, not skate, on the slick, shiny ice of the runway. I scanned the scene of Hercs, trucks, and tractors moving about. Parked near us, other Hercs sat; one powered up its engines and started to taxi for takeoff—to where, I wondered. New Zealand, the South Pole? Vapors with an exhaust smell exited tailpipes of trucks and vans that rolled by. Men moved about and did the jobs needed at a landing field. Ice isn't tarmac, but all activity looked normal for an airfield, and it was occurring on ice floating on the sea. That captured my imagination for a moment, and I reflected on the hubris of the human effort in Antarctica. Who comes up with the idea of an airfield on sea ice, and then builds it? That appeared to me an effort to control Antarctica, to challenge it. Soon, I'd encounter many examples of the failure of that folly.

Welcome storm in McMurdo. (Photo: Chris Siddoway)

Dave, Steve, Tucker, and I mingled on ice the color of slate, exchanged high-fives, giddy that we'd arrived at the end of the world. Most words said to each other got drowned out by the volume of our Hercules' engine at idle. "I can't believe we're here," I yelled to Dave. Lightness and a rush overcame me.

"Man! Yeah!" Dave had a wide grin, shook his head side to side. His face showed a laugh I couldn't hear. He shuffled in a circle, his large shoulders hunched only a bit—his characteristic gait. We assessed the alien landscape that opened before us. I absorbed the scene: overcast with a slight breeze. I turned into the wind. My face stung from the cold. I expected that, so I closed my eyes and let it touch me like a welcome kiss.

The scope of our enterprise took hold of me. I absorbed the bleakness. The whole idea of what we would soon do—work eight hundred miles from McMurdo—struck me as extreme. *I can do this.* What did Dave think? Tucker? I assumed Steve thought about getting to work.

We'd landed on the Sea Ice Runway, on about six feet of ice floating above many hundreds of feet of ocean, right there in McMurdo Sound. Miles of ice stretched in front of us, reaching toward the base of the Transantarctic Mountains, a hazy steel outline in the distant west. In the opposite direction, at the edge of Ross Island, a few dozen drab buildings with muted colors sat spread over a patch of black hillside. These structures rested at the foot of a giant white mountain, which emitted a stream of vapor that trailed sideways in the clouded sky. McMurdo Station lay on a black lower skirt of Mount Erebus, a thirteen-thousand-foot-high volcano that showed magnificent indifference to the puny, helpless settlement before it. *Why build it here? That's an active volcano. Doesn't that matter?*

I glanced around for clues about what would happen next, paced back and forth. In a few minutes, a truck fitted with a large ashen plywood box, like a small moving van, pulled up to us. A man in a pale brown Carhartt parka and overalls opened a rear door on the box. Without expression, he directed us along with a group of other passengers to load up, get in, and sit on wooden benches along the sides. I gave our team a follow me nod. Awkward in our bulky ECW, we struggled up the few chest-high metal steps, passed our clumsy orange bags inside the box, got in, and sat on a bench with a lumpy pad. Moving around in these clothes took effort and many deep breaths of cold, dry air.

The man in the brown parka shut the door to the box and latched it. No windows. Dark and cold but quiet and out of the sharp breeze. My breath steamed. This truck couldn't compare to a Welcome Wagon. I gave up thinking I could control our circumstances. He started the truck, drove for about five minutes across the smooth sea ice, then slowed and began to lurch and wobble the truck across the Transition, where the ice and solid ground of Ross Island met, before starting uphill. The truck continued up, gained elevation on a road I couldn't see, and made a few turns. He stopped the truck, opened the door to the box, and we piled out into the frigid, drab air.

Our group of just-arrived passengers now stood before the NSF Chalet, a quaint A-frame structure that reminded me of a down-market ski resort. The Chalet overlooked the frozen sound and Sea Ice Runway. Flags flew from a set of poles behind the building. They stood sideways, as if starched, held up at attention by the cold breeze—twelve flags of the original Antarctic Treaty nations. To my left stood a brown wooden two-story, peaked-roof building. The front side displayed a sign that read Mammoth Mountain Inn, yellow letters on a blue background. I said to Tucker, "Look, someone has a sense of humor." Under his parka hood, he raised his eyebrows. A man behind me pointed to the next building down—two stories, metal sides, industrial looking.

"That's Hotel California. These are dorms, for transients," he said. *Hey, that's kind of cool. Are we transients? Could we check out any time we like, but never leave?*

A short distance behind those buildings, a brown-black cone-shaped hill with a cross on top defined the horizon. A cross—a memorial. A sense of foreboding rose in me. I swung my arms and stomped my feet. I elbowed Tucker and pointed to the hill with the sprouted cross.

Whop-whop-whop-whop-whop. The noise got louder as a helicopter approached behind the dorms, a Huey like those flown in Vietnam, but bright red, with ANTARCTICA in tall black letters on its side, NAVY on its tail. Would it land? Where? The helicopter hovered, dropped low, disappeared behind the dorms; its engines wound down. Soon, another helo powered up. The noise increased; it appeared above the dorm roof, then turned to head across the sound. Who could sleep in there?

I looked about in different directions. Snow cover appeared sparse in town, with some mud in places, this late in the day. Small rivulets cut across the roads. The black ground absorbed heat from the filtered late-spring sunlight and melted the snow.

We made our way inside the Chalet for an in-brief. The main room opened to a high roof; colorful international flags grew out from one brown wooden wall. Metal and wooden plaques covered another wall, awards of some type. Large, framed maps and satellite images of the region occupied another wall space. The room's warmth announced a misleading notion of civilization. The staff had arranged metal folding chairs to hold about thirty of us. We faced toward McMurdo Sound but couldn't see it; windows high up in the roof let in soft light. I wondered why they built that wall and blocked the spectacular view. People maneuvered to sit in the small chairs wearing their bulky Big Reds, slung them over the chairs, or lay them on the wooden floor.

The NSF station leader came out to greet us. Erick. He wore jeans and a down vest, an L.L. Bean look. Erick appeared young, but when he entered, the room became quiet. He took charge. Three attractive young women came out of a side office to pay attention to him. His staff? Attired like Erick, they scanned the room. I watched them for a minute, searched, and noticed they all had wedding rings. Any husbands here? Did they leave them in the US while they spent months in Antarctica with hundreds of horny men? Would I agree if I had a wife who wanted to do that? Doubt it. Maybe that's why I couldn't stay married.

Erick began. He gave us answers to obvious "what now" questions. The station would be full soon, almost a thousand people. He told us the three weather classifications in McMurdo that we'd find posted around the station and on TV scrolls in the dining hall, called the Galley, and dorms:

- Condition Three is normal, like today; travel and activities are not restricted. It was +19°F now, but with the breeze, the wind chill read between +6° and +9°F.

I knew my Big Red and I could handle that.

- Condition Two is when winds hit forty-eight knots or higher: that's almost sixty miles per hour, and wind chill −75°F or below. For Condition Two, check in with the Firehouse if you need to travel.

What? I wouldn't be going anywhere.

- Condition One is wind above fifty-five knots (sixty-six mph), wind chill below −100°F, visibility less than one hundred feet. Stay inside is the rule for Condition One.

I worried about our preparation; if Condition One hit us, would we and our tents handle it?

He told us where we would sleep, eat, find a doctor. To make a phone call, we'd go to the New Zealand Scott Base over the hill. We learned where to pick up mail, a couple of times a week, maybe; where to get a drink (four bars in McMurdo); where they allowed you to hike; where not to go. Everyone attended safety briefings. Everybody went to Survival School. He said that a few years back, two guys fell into a crevasse behind the station and got killed. Each team would have a science team meeting and needed to prepare for work. He fired off much more info than I could absorb. *Can I read this somewhere?* Chris and Cain had worked here a week. I counted on them to know these details.

A second man spoke next—older, gray with a beard. He wore a red-black plaid USAP-issue wool shirt and black wind pants, not outdoorsy street clothes like REI Erick and the women staffers. He didn't act friendly—no smile, not relaxed, a hard-ass. He told us no drugs allowed, or you're outta here. He looked around the room, checked if we were on receive. I knew this, muttered to a guy next to me, "What about the dog at Christchurch CDC that sniffed us down. Doesn't that count?"

"There're ways around that," he whispered, smiled like a wink at me. An OAE. He must know the secrets. I wondered about the workaround—but wasn't interested in dope. The red-plaid man said federal criminal laws apply to everyone. He identified himself as a deputy US marshal. I wanted to ask if the station had a jail. Any guns? I tried to imagine how an arrest would play out.

Erick came back and described what teams needed to do if they were going to the Pole—he meant the South Pole—or going to the Dry Valleys, or working in labs in McMurdo. He told winter-overs what they needed to do. I glanced sideways around the room, tried to identify the weirdos who'd volunteered for that nightmare. Erick finished without mention of Marie Byrd Land. I turned to Tucker, "Doesn't he know about us?" He shrugged. I got it; we had a fringe project, and nobody went to MBL. We couldn't fall through the cracks in this system, get ignored, the short end of the stick. I needed to watch out for that.

Dave, Steve, Tucker, and I got our dorm assignments. We would be together. Later I'd need to find Chris and Cain and get their updates. We exited the Chalet and began to make our way to our dorm a few hundred yards in the distance, on the other side of town: Mac Town (nobody said McMurdo). "I guess we'll walk," I said. My gear and clothes weighed at least seventy pounds.

We slipped on our rucksacks, zipped up our Big Reds, and grabbed our orange clothing bags, one over each shoulder. I clenched my teeth, dropped my head, and got underway.

The four of us lumbered along, faced into the cold breeze, carrying and then dragging our bags in the black dirt and cold mud. I huffed and puffed; my Bunny Boots reached up to my calves. Just to walk took hard work. *These boots are made for walkin'*... I hummed to myself, but these boots weren't made for any human activity. I checked for a wheeze—not sure, but I caught my breath. *I'll ignore that until I know it's my asthma.*

The legs of my long john pants twisted into tight vises around my calves. Inside the boots, my socks had come loose. I had to stop and adjust my pants every twenty or thirty yards. Nobody else did. Wrong size boots?

The place looked almost familiar, a mining camp in the Arctic or a drab, decayed Soviet town in Siberia, or a mix between them and a space station from a horror movie. No detectable charm, but I didn't expect warm and cozy. A few low brown-black volcanic hills formed the backdrop to Mac Town; Mount Erebus behind them boasted its presence. My geologist brain asked me what caused that monster to form here and when had it last erupted? For sure if it blew, Mac Town would be buried.

Large white cylindrical fuel-storage tanks dotted black hills right above us. Arrays of thick pipes ran parallel to roads, one pipe labeled JP-8, jet fuel. If a truck ran into one of those, leaked fuel could wash downhill into the roads and ditches. Atop one black hill, a giant white golf ball maybe thirty feet across sat teed-up—radar, I guessed.

The roads cut into volcanic rock and ash. Trucks, vans, and heavy equipment moved around, stirred up dust; the breeze carried clouds of fine coal-black dirt everywhere. Erick's briefing had told us to be careful of traffic if we had our parka hoods up with face tunnels extended. We wouldn't be able to see or hear vehicles. True: a front loader with aggressive tires as tall as me scurried by my side—I didn't see or hear it approach.

We walked through a mix of stark military buildings—lead-colored and rectangular. Quonset huts that looked like coffee cans cut in half lengthwise, metal warehouses, plywood shacks, and Jamesways—Quonset huts made of canvas—appeared here and there. No zoning rules in effect. In the distance, I could see tall, dark brown buildings: three-story dorms. We'd stay in these, not the transient dorms by the helo pad. A good start—I hoped they'd be comfortable.

We passed by the Galley and mess halls in Building 155, the town center. A mustard yellow, two-story structure with metal sides and roof, and rare small windows. The building covered maybe a half-acre. On the dirt road behind 155 sat dumpsters. Skuas, large dull-brown and belligerent Antarctic versions of sea gulls, raided the bins, busy with their dinner, shouting caws, spreading trash all over the place. Survivors. Bits of paper and cardboard blew in swirls of man-made snow down the road. I didn't expect Club Med, but why make this mess? No vegetation. The Antarctic Treaty forbids living plants here because they could start a new outlaw ecosystem.

A hydroponic greenhouse lay on the edge of the cluster of Mac Town structures—an exception to the rules where meager amounts of fresh greens, or "freshies," grew inside a plywood garage under floodlights. Rumors drifted around that pot grew in there: some seeds smuggled south, plants hidden. I wondered again how dope got by the sniffer dogs.

We arrived at our dorm building. I puffed and struggled up the few metal stairs, pulled the door open against the wind—much stronger on this side of town. We entered a vestibule, then a narrow and murky hallway with brown indoor-outdoor carpet. A small window at the far end let in a dull glow of frozen light. Relieved to be out of the frigid wind, I couldn't feel the tip of my nose. My face hurt. That hike carrying my bags in heavy cold weather gear had taken a toll. I caught my breath. "Shit, man," I struggled to say to Dave. He gave me a questioning look. I'd trained with a workout program at the Y. Guess I left out some key routines.

The dorm had two-person rooms that shared a bath. Our room had a dark brown utility carpet and coffee-colored walls, sturdy natural wooden twin beds, and a built-in wooden dresser. "Hey, this isn't bad. I could stay here a while," I said to Dave, then remembered I wouldn't. In a few days, our team of six would live in tents in the lonely wilderness and work in weather just like we'd left outside the door.

We unpacked a bit, then stretched out on our bunks to regroup. I needed that.

"We did it, Dave. We made it to Antarctica."

"Man, can you believe it? Crazy! Ha!" He kicked off his Bunny Boots. "Hey, look at us. We're here!" I could hear him laugh now.

"We're gonna have some great stories and do great science. We're going to uncover new ideas."

"Hell yes."

My asthma made me cough, more than once.

"You okay, Bruce? You seem a bit winded."

CHAPTER 4

The Lions of Antarctica

It is better to go skiing and think of God,
than go to church and think of sport.

—**FRIDTJOF NANSEN**, Polar explorer

McMurdo Station, November 21, 1989

Dave, Tucker, Steve, and I made our way to the dorm exit. Wind pushed the door shut. I leaned into the door, slipped through, and stepped out. Sharp cold slapped me silly. Nasty. Yesterday I had worn a T-shirt in Christchurch. Time for dinner in Building 155. The Galley stood across a dull brown dirt plaza, Derelict Junction. A few people strolled over toward 155, not in a hurry. They were dressed light, in thin windbreakers and jeans. I supposed I would get adjusted, but I cinched up my parka hood.

I looked toward McMurdo Sound and the Sea Ice Runway—a wall of white. Clouds had lowered and thickened. A heavy snow fell. White flakes in wild motion obscured my view of the sea ice and sitting flock of Hercs. Our arrival flight had lucked out, landed ahead of a whiteout. Luck mattered in a place where everything was out of control.

Transantarctic Mountains seen from Hercules.

At the entrance doors to 155, through a vestibule, I read a sign: "Attention: Please remove ice protection shoe devices before entering Building 155." Really? Folks forgot to take off their crampons? Bet they drove their moms crazy. The main corridor, wide and long, stretched ahead, full of people. A few dozen, mostly men, moved along. They came toward us and went with us. Boots sounded a roll of bumps on the linoleum floors; fluorescent lights showed military-green walls, notices posted along them. This was the Highway One we'd heard about.

Contractors with Antarctic Services—ANS—wore brown Carhartt work coats or red parkas. Their garments showed use—dirt, grease, and worn spots. Navy seamen and aviators wore olive parkas and jump suits. Big Red parkas announced the Beakers. Our parkas looked new, clean red. We got a few curious glances, but most folks were not interested. I felt self-conscious, tried to avoid eye contact. "Can't hide that we're FNGs," I said to Dave. He smirked.

Faces and hands poked out from the ANS brown and red parkas—hands like farmers', worn and beaten up. Most men wore scraggly beards. Men and women had raccoon eyes from the wind and sun. Tucker walked next to me. "These people look tough," he said.

We entered the Galley through a dark corridor. On the right wall, red parkas hung on hooks. On the left wall hung black-and-white photos of men with frozen beards; it looked like early days, and harsh times.

A door on the left opened into a large mess hall. I looked in, saw navy kids and older men who ate at long tables. We continued to move forward, following the navy officers, in their crisp olive-green uniforms and perfect haircuts, and the Beakers. I figured out that the Galley separated scientists and navy officers in one room, enlisted men and ANS contractors in the other.

"Look how everybody's segregated. That's bullshit," I said to Tucker.

"Huh." Tucker looked around like he hadn't noticed. "Guess we'll need to go along with it."

A cafeteria line formed. Tucker and I grabbed trays and looked at the menu set before us. I smelled fried food. "Hey, mushy veggies, fried stuff, salt and fat for dinner," I said to Dave, amused at the grossness of it. But I did see some freshies that made the flight south with us. Later, I'd learn any arrival of freshies would make big news in Mac Town.

Our trays loaded with food, we entered the main mess hall, smaller than the other room. No windows in it either. Separate tables for maybe six or eight people each. The room had the hum of chatter. No view. That was too bad. The navy officers had their own tables. The aviators kept to themselves too. This place was stratified—that disturbed me. I noticed a table of women navy aircrew. That was cool.

We saw Chris and Cain across the room. I recognized Chris' trim, athletic form and blonde hair. She grinned at us. Cain stood erect, about the same height as her, his angular face clean-shaven, his light hair thin on top. She waved, and we made our way to their table. Chris wore the Patagonia clothes we'd bought for the team—a new Capilene pullover and sleek, pale blue wind pants. Cain wore the issued ECW. Both of their faces were imprinted with wind and sun.

After hugs, handshakes, and a few backslaps, we sat down to catch up. We made small talk for a bit—how many days it took to get a flight south from Christchurch, the weather in Mac Town. The long flight and hauling of gear had tired me. I hoped they'd give me good news that everything was going smoothly.

"How's the preparation going? Any snags?" I took a bite of salty stew and sticky, bland white rice. Chris sat upright, elbows on the table. I felt energy radiate from her. She went over a list of tasks they'd done last week and what we had to do next. She named all the facilities they had contacted and mentioned ANS people by their first names. It became clear to me that she liked logistics and had it in hand.

"So, Bruce, we designed and built platforms to stand on and upright handlebars at the rear of two sledges," Chris said. Her grin told me she thought this was an essential accomplishment.

"And added footbrakes for riders to control downhill speed when sledges are towed," Cain finished her sentence. Brilliant. I never would have thought to do that. I felt reassured. These two were a good team.

"Yeah, and Bruce, the navy air squadron, VXE-6, told us they would Put-In sixteen thousand pounds for us, just one flight," Chris said. I couldn't tell if she thought this was good or bad.

"But Hercs can carry twice that, easy. Do we need more?" I said, wary that a problem was surfacing.

"They won't load that heavy, land on a snowfield they haven't been to before. Don't want to drop a ski into a crevasse, get stuck there," she said with authority. Okay, I got that. Chris told us how she and Cain had estimated weights of fuel, food, supplies, and equipment—they figured we could do it.

"What about our Pull-Out weight? We'll collect a few thousand pounds of rock samples," I said, and looked at Dave. His rock chunks for age dating would be the size of basketballs.

Chris told me the rock load would be offset by the fuel and food we used over six weeks, but she'd asked for two Pull-Out flights to cover us. For a moment, I considered probing her plan, but I felt pretty good about her news. I smiled, glanced around the table, and saw the team nod in agreement.

"We have our science meeting with NSF and ANS tomorrow. We'll learn more about what VXE-6 is willing to do," Chris said.

"We gotta plan and get ready for our shakedown trip," Cain added. A shakedown trip is a dry run to make sure all your equipment is necessary and in working order.

"We think we can travel over the sea ice north to the huts, to Scott's and Shackleton's," Chris said. Now she had my riveted attention.

"That would be so cool," I said. I could visit the source of my Antarctic obsession—where my idol Ernest Shackleton had lived in 1907.

"Should be just one overnight, twenty miles each way, most of the route marked by flags," she said. "We'll check out all the camping and travel gear, our snowmobiles." I was relieved they had this trip planned out. Even though I was the expedition leader, they had used their advance arrival time to get oriented. I was glad they hadn't waited for me to direct them.

"That's great, Chris. So, what's the tent situation? Do we get an extra one for a woman?" I asked.

"I don't want to tent alone. I'll share," she said. I knew she had shared a hotel room with Tucker in Washington, D.C., when we attended our orientation conference, so this made sense. She must not be bashful. Maybe Steve would be a practical tentmate; they shared the same research objectives.

"Well," I said, "Dave and I should be tentmates, since we're responsible for planning, and Cain and Tucker should share, collaborate on travel and safety," I said.

"Great with me," she said. "Steve and I are both working on the metamorphic rocks so it would be good for us to tent together."

"Everyone okay with this?" I asked. My tone said I decided it was. Nods all around the table; it was.

The conversation taught me who had a grip on our program—my new graduate student, Christine. I thought back to our first meeting. A couple of years back, I had distributed a blue memo to grad students looking for a candidate for this project. It had been in mailboxes for only a few hours when she knocked on my door. I recalled that meeting, learning of her geologic experience in the Arizona mountains and that she had a master's degree. More than that, she grew up on a small farm in Wisconsin where one of her chores was feeding the calves before dawn. I didn't know her very well yet, but I liked her take-charge, can-do attitude. I would be relying upon my whole team, not only to do their specific work, but also to keep us all safe.

We were after big questions. Everyone except me was experienced in wilderness work, and they were smart, independent thinkers. Without prior expeditions to Antarctica, I felt vulnerable now that I was here, facing the reality of the harsh conditions and that we would be completely on our own and far from rescue, should we need that. Leading this team would be a challenge. I was accountable as the leader, the PI, the principal investigator. I needed to understand what they wanted and make sure it met our objectives. For a moment, I felt at the deep end of the pool, trying to stay on top wearing soggy ECW.

A lot of pieces were in motion now: planning for food, equipment, and supplies; preparing for Put-In; training for wilderness work. Then there was the weather. From what I'd heard, it could be like nothing I'd ever experienced or imagined.

* * *

As Dave and I entered our dorm, I shook off my tension along with my thoughts. I told myself what I needed to believe: *we'll be fine.*

"I'm going to check out one of these bars," I said to Dave.

"Which of the four?" he asked, laughed, and shook his head. I was surprised too that there were so many places to get wasted here, but it made sense. I imagined a few months working in Mac Town could be a burden. Chris had told us the O club, the officers' club, had the best vibe; that's where scientists and officers hung out. More segregation. That worried me. My experience on research ships taught me that segregation of the workforce spurred resentments.

"Officers' club tonight," I said.

"I'll catch up later, maybe. I'm beat, man."

Outside, the snow had stopped; it looked like the sky would clear. I headed in the direction of the O club, used my Mac Town map, and had no idea what to look for. I saw a rust-colored pair of Quonset huts joined at right angles, like a T. A few people entered a weather-beaten wooden vestibule at one side of the huts.

In the entrance, Big Reds hung on hooks attached to a plywood wall. I walked into the shadowy bar and lounge—no windows, the walls paneled in what looked like varnished cedar. Skis and sledges hung from the half-pipe ceiling. Overstuffed sofas and chairs sat here and there along with a few low tables. Music played, maybe Judy Collins? Wine, beer, and scotch, each a dollar, served at a wooden bar with a polished top and tall captains' chairs. Hey, all this was primo, civilized. I felt good, upbeat. I looked around and saw a geologist I'd met in Christchurch, Mark.

We started talking shop. He covered their fieldwork plan for the Dry Valleys. With no snow in the valleys for sledge travel, his team planned to hike to all outcrops. "But we get helo support," Mark said, "and can cover lots of elevation changes without climbing." I felt jealous to hear it—helicopter trips around the Dry Valleys, the most dramatic scenery on the continent or maybe on Earth. Our team would work way beyond helo range—no air taxi service and no helo rescue.

"Remind me what you're doin' here, Bruce," Mark said.

"Headed to the Ford Ranges in Marie Byrd Land, MBL," I said.

"What's your project there?" Mark asked, determined to move deeper into science.

"Work in the Fosdick Mountains in the Ford Ranges mainly; rocks there might match up with rocks in New Zealand," I said. "Find the story of Gondwana breakup, the timing and effects.

"I'm doing paleomagnetism of rocks, looking for evidence of tectonic movements, like spreading between East and West Antarctica, and separation of New Zealand and its submarine plateaus from MBL."

"Am I remembering hearing about some work in the Ford Ranges?" Mark asked.

"Yeah. A few people went there years ago but didn't go back." I realized a message resided in that fact, but what was it? "A US party in the mid-sixties spent a few weeks there, and a Kiwi party in the seventies, but they didn't get as far out as we're going."

Mark nodded to signal he followed me.

"You gonna do a Herc Put-In, Deep Field?" he said.

"Yep." I let myself smile. I'd have fun bragging about that when it was over. The bigger payoff would be finding a connection between the geology in New Zealand and the Fosdick Mountains—if one was there. That finding would be pure elegance, solving how Gondwana looked before it broke apart and what happened next.

We drank more, and I talked about how my team would conduct our work, blaze trail, live out in the wilderness, and so forth.

This had been fun, but with the time near midnight, I had things to do tomorrow. I said goodnight and made my way out the door. "Ah, ah, ah…" The light blinded me. Then I remembered there was no sunset in late November at seventy-eight degrees south. I had forgotten, or maybe I was too buzzed. I dropped my head, squinted, and dug into my parka pockets for sunglasses—which pocket? Too many of them.

* * *

The midnight sky showed itself calm and clear. With great weather now, I decided to take a stroll. I faced left down a black dirt road toward McMurdo Sound and saw a small white church at the end of the road. *What*? Small, like a shrunken-down version of a New England church, complete with a steeple and bell. The building sat at the edge of the sound above a slope that had a good vantage point.

At the church, a sign read "Chapel of the Snows"; I went in through an unlocked door. I figured it was always open for business. It had been a couple of decades since I'd been in a church—I'd decided I'd had enough of Confession. Didn't look too much like a church inside, with minimal, plain walls. No one here after midnight—quiet, nice. I felt comfort, took a few slow, deep breaths. Rows of padded chairs faced an altar table backed by a stained-glass window. A nice touch, but it blocked the view.

A room to the right had windows that looked over the frozen sound. In the room, a coffee urn emitted a welcome aroma. I grabbed a cup and stood to look out, my first chance to see it all and think. Alone.

The scene presented white and deep brilliant blue. I took in the raw beauty, focused on it. The sun sat low in the southern sky. I looked west, miles across the ice-covered sound—an immensity of ice. The scale of the view threw me off balance. Only two weeks ago, I'd arrived in Christchurch. I'd wandered into a park that followed a river and leaned over the rail on a cobblestone bridge to watch punters in white outfits and straw boaters pole tourists along the narrow course of the Avon River. That waterway wound in lazy loops through downtown Christchurch. Distinct from River Avon of Shakespeare fame, I nevertheless felt Anglican charm. The green lawns bordering the channel and the willows that drooped down to touch the river's surface had caused me to wonder if a cold, barren place like Antarctica existed on the same planet and could be reached by a flight south.

At the far western extension of the sea ice rose the carved, faceted peaks of the Royal Society Range. They glowed white. Shadows of the peaks lay over wide glaciers that cut through them on their route to the sound. I knew these mountains stood thirteen thousand feet above the sea. I could make out horizontal rock layers in the higher peaks. I realized those layers must have been formed under the sea, then uplifted to form the foundation of these mountains. When did that happen and what was the cause?

I imagined what it would be like to stand on those mountaintops, to view the infinity of ice beyond them, the excitement of that, the aloneness. The longer I stared at the range, the more my imagination worked. I saw the peaks as members of a pride of white lions—they crouched at rest. They faced me, sphinx-like, guarded the colossus of the interior. They dared me to enter.

Soon we would find ourselves in the wilderness—in conditions much harsher than here. I'd have to be ready. "Are you up for this?" That's what the NSF official asked me in an exciting phone call when he told me our project was approved.

"It's not like you'll be working in the Transantarctics, or the Dry Valleys near McMurdo," he said. "You'll be way out there. No rescue by helicopters—too far. You're on your own." He waited a few seconds for that message to set in. "Marie Byrd Land has a nickname. You've heard it, right? Mighty Bad Land."

CHAPTER 5

Radio Room

This continent is the highest, driest, coldest, windiest,
emptiest (and silliest) place on Earth.

—From **US ANTARCTIC PROGRAM BROCHURE,**
amended by an Antarctic geoscientist

McMurdo Station, November 22, 1989

"You'll need ten days or more to get ready," Erick the NSF station boss told me. I felt on overload. Never mind that those ten days sounded too meager to prepare for an expedition to MBL—the wilds of Antarctica. Our team sat at a table in the Chalet with NSF officials and the ANS staff responsible for our Put-In to the northern Ford Ranges. I tried to take notes but couldn't keep up with the high information rate. I decided to rely on Chris, who knew everybody at the table—and in town. She knew what we needed to do before anyone said it. Erick said that there were eleven remote field parties planned for the season.

"Most are close, in the Dry Valleys, in helo range. A few are way out in the Deep Field like you folks," he said. "In fact, the WAVE team will be the farthest. A Herc needs to refuel at Byrd Surface Camp to get to their work area."

"Just the WAVE team and us out in MBL?" I asked.

"That's right. So, you're project 070, but I hear you use an acronym, FORCE—is that correct?"

"Yeah. Ford Ranges Crustal Exploration."

Someone at the table muttered, "May the Force be with you." A couple others sniffed a laugh.

Comms, or radio training, would happen today. In the days that followed, we would learn how to operate and repair heavy-duty snowmobiles; plan, gather, and stage equipment and food at the Berg Field Center (BFC); plan and organize our loads for the Herc; make a shakedown camping trip; take Survival School and learn crevasse rescue; make a reconnaissance, or Recce flight, then get dropped off—the Put-In. I wondered why we would make a shakedown trip before Survival School. What if that trip turned into a survival situation? I didn't ask—might have a simple answer. This wasn't the time or place to act stupid, to act amateur. I felt under a microscope.

Chris already had an idea of how much equipment and supplies we needed. She told us we'd need thirty wooden rock boxes for her and Dave: rectangular wooden crates eighteen inches wide on the long side, a foot square on the ends with rope handles. Loaded, they held almost one hundred pounds. She and Cain calculated the fuel we needed for snowmobiles and stoves, and our food rations. They estimated the total Put-In load would be sixteen thousand pounds.

"That means two Put-In flights," Rick, the thirty-something ANS fixed-wing flight coordinator, said. We'd visited his office earlier in the corner of a beat-up wooden shack, walls covered with posters of motorcycles. He chain-smoked, leaned back in his chair in the relaxed manner of a person who knows their business, and adjusted his thick-lensed glasses and black leather jacket that he never took off, at least any time I saw him.

Rick told us that the Hercs could carry more than sixteen thousand pounds, but wouldn't carry that much to land in an open snowfield where no one had landed before. Just like Chris had told me earlier, "Lighten the load by airdropping fuel barrels ahead of Put-In. Divide your loads into two flights," he said. The plan would be for a second Put-In flight on the same day as the first. "Make the first load for essentials in case the second one can't get out there for a while." Two flights. I rubbed the back of my neck, which had stiffened. Might be a fuck-up and only one flight; that's what he'd just told us.

* * *

I wanted to know about the radios, how to communicate with McMurdo—the bottom line to our safety. Our team left the Chalet, made our way across the black-and-gray volcanic dirt road to building 165, a two-story, mustard-colored metal building with antennas and radar domes sprouting from its roof. We clumped up the metal stairs and opened a beaten, heavy weather door. The inside reminded me of navy ships I'd been on—plain gray walls, the smell of floor cleaner on linoleum, fluorescent overhead lights; men in olive-green fatigues moving about in the corridors.

We met a burly but nerdy navy non-commissioned officer—NCO—a radioman, in his office. He took us to a small windowless room with radios stashed on metal shelves along the walls. We crowded around a table with him.

"Where you guys goin'?"

"Ford Ranges, Marie Byrd Land," I said.

"Never heard of them. Near Byrd Station?"

"Almost. A few hundred miles short but eight hundred from here."

Never heard of them?

"Well, no matter, you're gettin' two Southcom 130 field radios. One is backup." He pulled two off the shelves and put them on the table. About the size of a medium attaché case, olive drab, twenty pounds each, with a distinct no-frills, rugged military look to them. Knobs with cryptic labels in front and several connectors in the back—looked complicated.

"What about walkie-talkies?" I asked. I was concerned about us separating while we worked.

"Sure, we have 'em—hard to keep charged up," he said.

"We don't need those," Cain said. "More stuff."

Surprised, I let his comment slide. We planned to carry police whistles I brought to signal each other; these would have to do.

The NCO started to give us a lecture on the radios, so I took out my field notebook. He told us we would use these to check in each day, report our status and our weather observations. We would learn how to do that with Mac Weather upstairs in this building, so he advised us to set up a visit. He grinned, maybe with the knowledge that we would have plenty of weather to report from Marie Byrd Land. He handled the radio, showing us the knobs on its front, then lectured us on antenna setup, and keeping the batteries charged, and the frequencies to use. "Ones you care about are eight-niner-niner-seven for Mac Center aircraft and weather and eleven-five-five-three for field party check-in, but that one doesn't always work," he said. *Great.*

He reminded us that we needed to check in with Mac Center daily, and if they didn't hear from us for three days, they would launch a SAR operation—Search and Rescue—and send a team by Herc out to find us. It would later turn out that on one day in the mountains, our radio didn't work, couldn't receive or send, and the backup radio didn't work either. Did both have shot batteries—frozen? We had three days to figure it out before SAR would fly out. They would find us alive if they did, of course, and we would be the butt of endless, merciless jokes in Mac Town— better off dead. Steve had used a voltmeter I brought and found a short in the antenna connector, bent by wind likely; we needed to resolder it. He had a one-shot soldering iron but no solder—why not? My voltmeter had solder globs on its circuit board. So, Steve melted solder off the voltmeter board and used that solder to repair the antenna connector—back in business in less than one day, no SAR.

The NCO paused and looked at us around the table, checking for questions. He'd dished out a lot of info. I hoped someone else had paid attention. I glanced sideways at the rest of the team. No one had taken notes other than me, but Steve nodded as the radioman spoke. *Right*, I thought, *Steve has a degree in engineering. We got this covered.* I relaxed my shoulders.

Dave and I had clashed about Steve coming on the expedition. That started back at UCSB when I told Dave about the phone call from NSF— the call that told me our project was approved.

"Yes! That's so great!" Dave said, jumping up and showing a wide grin. We talked about the particulars of startup, in a fuzzy way—didn't know enough yet. I left to tell the front office the news.

An hour later, Dave found me.

"I spoke to Steve, asked if he wants in, and he said sounds good," he said.

"What? What do you mean?" Red warning lights flashed in my mind. Dave had run off and acted on his own. Would we butt heads on this project? Conflict now before we picked the team—and we weren't even there?

"Steve Richard. We need a structural geologist."

"Hey, slow down—we're partners in this. Don't go recruiting for our team unless you and I talk it over." I set my jaw, acted pissed. "We're looking for folks to commit for a few years. They must be up for it. Have to be able to deliver, got to be the best."

"You're right, I got ahead of myself."

I quickly thought it over. Steve would be perfect. Dave had made a good call, but we couldn't make this level of decision independently, without agreement.

"Yeah, Steve is the best." I had made my point. "Let's go see him."

Steve and Dave were peers and at least ten years junior to me. Both recent PhDs from UC Santa Barbara, they had worked together as grad students. We found Steve in his office in front of his Macintosh, a rendering of the Colorado desert on the screen cut into pieces he could move around—to restore fault slip and crustal block rotations. That new, quantitative method fit with his former life as an engineer. After his bachelor's, he discovered that he wanted to explore the Earth, have more fun than at a desk and lab bench, so he changed his path to geology. Not as tall as Dave, thin, wiry, with a scraggly black beard. A cheerful man, thoughtful and congenial. He and his girlfriend loved contra dancing. He would be valuable company.

41

We pulled up chairs, sat down to share the thrill of this news. He asked questions to see if the project intrigued him and fit his geologic curiosity. After we explained, he said, "Ha, now that's cool. Sure. I'm in."

"So, you guys got this?" the radioman asked.

You're kidding. No book on this?

"We want to see the comms room—see who we will communicate with," I said to the radioman.

"Okay, sure," he said. "The guys who you'll talk to are upstairs." I nodded affirmation. He told us to take all this gear with us. I picked up one radio in a canvas bag and slung it over my shoulder; Steve and Tucker got the rest.

He led us out the door and up the stairs to a room with a Communications sign on the door. *So, this is the money room*, I thought, *the nerve center.*

Inside, the sounds of loud rock music hit me. Metallica, Megadeth, or Black Sabbath played at full volume. Guitars and vocals screeched from large box speakers. *What the hell's going on here?* Electronics, lights, dials, and gauges covered one wall. Four, maybe five kids in navy fatigues sat at a long table in front of a wall of radio equipment—skinny, some with acne, wearing headsets. *Is that why the music's so loud, so they can hear it with headsets on?*

The radioman, unfazed by the music, introduced us to the NCO in charge of this chaos, a comms guy who looked in most ways the same as the radioman. The comms man started to explain. I only heard a bit of what he said—the crazy music was so loud. I heard him say, "aircraft comms are here and South Pole here," and so on. "Stay off eight-niner-niner-seven, the Herc freq," he said.

I barged forward and shouted over the music. "Where're the field camp radio monitors?"

He motioned us to follow him into a dim back room. He pointed to a metal shelf with a radio a bit bigger than the one I carried.

"This here is where we monitor eleven-five-five-three for field camps," he told us. I could just about hear some scratchy chatter coming out of the unit.

"But there isn't anyone back here," I said. I wanted to say: how the hell could they hear anyone call in over the heavy metal blasting in the front room? I started to protest but held back.

"Not to worry," the comms man said. "We check this here set every now and then." Every now and then? I stood silent and dumbstruck, not reassured, not certain what to say.

"Thanks." I led our group out. Down the steps in front of the building, I said to the team, "See what went on in there? Unbelievable. They care about Hercs and South Pole, don't give a shit about teams like us." A few of the team looked at me but didn't speak. I couldn't tolerate possible radio comms problems. My team needed that safety net.

"Know what?" I said. "Doesn't matter what they told us in there. We want something or we're in trouble, we call in using the Herc frequency. What are they goin' to do about it?" That made sense to me. I felt a bit calmer.

I slung the radio bag over my shoulder and hugged it close to me. We walked in the cold, bright air and drifting dust back to our dorm. The radio would go next to my bed, here in the dorm and later in my tent next to my head. *I'll know this radio inside and out. Now it's my new best friend.*

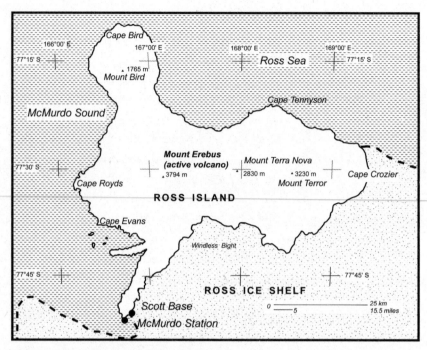

Map 3. *Ross Island in the Ross Sea, Antarctica. Places noted or mentioned in the story include McMurdo Station, Scott Base, Mount Erebus, Cape Evans (Scott's hut), and Cape Royds (Shackleton's hut). (From Wikimedia Commons)*

CHAPTER 6

Cape Evans

To strive, to seek, to find, and not to yield.

—**ALFRED, LORD TENNYSON,** *Ulysses*

Cape Evans, McMurdo Sound, November 23-24, 1989

I crawled out of my tent to take in the view to the west across the frozen
reaches of McMurdo Sound. The sharp white peaks of the Royal Society
Range stood upright, tens of miles away. Their magnificence captured
my dreams. The sun hung low in the sky—the time was near midnight.
I thought of my good fortune to experience the sight spread before me;
the privilege sank in. For a moment, I closed my eyes with that realization. The air felt calm and very cold. Not miserable, but it would be
with any breeze at all. At the edge of the sea ice a short distance down
the snowy hill from our camp, I studied a low wooden hut on the rocky
promontory of Cape Evans—Robert Falcon Scott's *Terra Nova* hut
where he'd prepared for his journey to the Pole in 1911. I located south,
to orient myself. The sun didn't set—it circled around in the sky, coun-
terclockwise, higher at noon in the northern sky, lower at midnight in
the southern. I stared south for a while, toward the Pole. I thought of
those who'd made their way into that infinite unknown eight decades

Traveling over the sea ice to Cape Royds.

earlier, on foot like Scott, or by dogsled for Norwegian polar explorer Roald Amundsen, nine hundred miles to the South Pole and then back. The magnitude of that journey stunned me every time it entered my memory. *Who thinks that way these days?* My eyes watered. Must be the cold air.

Dave and I had just finished our Thanksgiving dinner of bean burritos covered in tasty melted cheese. Thanksgiving, Christmas, New Year's, and my son's birthday would all be spent on Antarctic snow. That had its advantages, like simplicity, but the downside was hidden loneliness. Tonight, I'd sleep in a Scott Polar tent for the first time, a robust pyramid-shaped structure designed for two people and with a record of surviving Antarctic gales. Now on our shakedown trip, we needed to learn how to camp and live in the Antarctic alone—in tents. Living in tents, in Antarctica, might prove brutal. I would find out. The weather was perfect, cold for sure, but I didn't feel chilled with all my gear on. Our sleeping bags would be fine.

Cape Evans

On the shore of a frozen bay north of Scott's hut, I saw the Greenpeace station. A low, green motel-style building, erected here without the sanction of the US or New Zealand Antarctic Programs, who claimed they ran things in this part of Antarctica. Never mind that no nation could claim territory under the Antarctic Treaty. We had toured the station earlier that evening and met the skeleton crew that manned it for the last eleven months. Their purpose here was to watchdog the conduct of science activities, watch for environmental violations. A few months later, I would encounter Greenpeace again.

Our team arrived at the cape in the afternoon on four snowmobiles that towed four sledges. The chilly fifteen-mile trip took several hours. We were well north of McMurdo and out of range of hikers, skiers, and unauthorized visitors, except for Greenpeace. I drove one snowmobile; Cain, Tucker, and Dave drove the others. Steve and Christine stood and rode on the platforms at the rear of two handlebar sledges, like dog sledge drivers, but a snowmobile, not a dog team, pulled them. Dogs were no longer allowed in Antarctica.

Before the shakedown, we learned maintenance and a few simple repairs on the snowmobiles—became familiar with them. Not the sleek, low-slung machines that speed over slopes in ski areas, these were large orange brutes, more like small trucks. They each had a large plastic windshield and a cargo area of a few square feet behind the driver's bench seat, oriented front to back, with pull cords to start them—at times requiring many exhausting yanks.

Tucker had brought toys. "Here's a thumb-lever bicycle bell to put on the snowmobile handlebars, and Clarabell horns for the sledge riders to signal the drivers," he said, then laughed, squeezed the blue rubber bulb on a horn, and attached a bell to his machine. We found out that the wind and snowmobiles made too much noise for these to be of use, but the idea struck me as funky and funny, and I let myself chuckle.

"I also have a mascot for us," Tucker said, and pulled out of his pack a stuffed gray mouse a few inches high wearing a tiny green jacket and pointed red hat. "This is Skido," he said. "Sue gave it to me for good luck and safe passage." He put it on the dashboard of his snowmobile. Looked silly and fun; maybe it would work.

We had left McMurdo midday, headed north on the sea ice, following a flagged route. Cain took the lead. Eight-foot-tall bamboo poles stuck into holes drilled in the ice every quarter mile or so marked the trail. Red and green flags atop these stood sideways in the wind, a few torn to shreds. The weather had been breezy and cold with broken clouds. We drove; the wind pushed at our back. The snowmobiles and sledges slid and whipped around on the sea ice as we made our way. I didn't know better, but I felt uneasy, that we drove too fast. I ignored my concern.

We had a guest, Jill from the BFC, the supply center. Chris had arranged for her to come with us. She told me about Jill the day before we left.

"Oh, and Jill from the BFC wants to come with us," she said.

"Huh…what for? Why?" I asked.

"She wants to get out of Mac Town for a break." Travel of staff in McMurdo was tightly controlled. They could not get off station without a good work reason. NSF remained rattled by the two guys who died in a crevasse while on a hike just behind town a couple of years before. Those bodies were recovered—after much effort. NSF figured the safest approach: don't allow staff to do much other than work. Scientists had freedom to move; they could make up any excuse, and we did.

Jill impressed me as a nice person but extra baggage.

"But we don't have a space for her to ride," I said.

"She's okay to ride on a sledge. And she'll bring her own mountaineering tent."

"The sledges are piled high with our gear," I said, almost amused—or annoyed.

"She'll strap herself on top of the load," Chris said. "Remember, Bruce, she's in charge of our equipment, food, and supplies, all our stuff." Chris showed me a knowing smile. She meant that if Jill asked to come, we had to take her. Chris took charge but was covering for us. That was good thinking on her part. Still, this trip could prove risky for Jill.

So, Jill stretched herself out on top of one of the loaded sledges, laid flat on her stomach, and held on to the tie-down ropes. We sped across the sea ice in a line following Cain. I looked at Jill, spread-eagle on the sledge in front of me. The sledges skidded back and forth; they snaked across the ice, bumped over ridges. She hung on tight. This woman was desperate for change. Glad we could help her out.

We made it to Cape Evans late in the afternoon. Cain directed us to a spot to camp up the snowy hill behind Scott's hut. We began to set up our tents for the first time. Cain and Tucker set up their tent to show us how.

A Scott Polar tent, a large orange canvas umbrella—a pyramid—without a center pole, takes two people to carry. Four wooden poles, two inches thick and ten feet long, were sewn into each corner and joined at the top. Dave and I dragged ours to a spot about twenty yards from another tent, far enough for privacy but close enough to see our neighbor in a blizzard. I glanced at him, sought out an air of confidence.

"I'll grab these right two poles; you take the other two," I said. We stretched out the pairs of poles as the tent lay on the snow and made a triangle.

"Pick up," I said, and we pulled the four poles apart and set the tent upright to make a pyramid. Dave and I took shovels and made snow holes for each pole end and stretched the fabric tight. We pulled out the skirt that surrounded the tent bottom and staked it down with sections of one-inch pipe we drove into hard snow with our rock hammers. They rang out when struck, then quieted down to a thud as they were driven deeper. Before I'd set foot on wind-driven Antarctic snow, I'd never thought it would resemble concrete. Each hammer blow made me feel more secure, tighter. Guy ropes needed to be staked down on each side, and that required coordination—they needed to be pulled at the same time. Dave grabbed the ropes on one side, and I did the opposite.

"One, two, three, pull," I said, and we yanked the ropes tight on the right and left sides, then the front and back. That filled out the tent, gave it the shape of a pyramid atop a short, square, orange canvas box. I liked the look. I expected a canvas smell, but I couldn't detect any.

"I'll be outside man and you inside," I said. Dave crawled inside through the entrance tunnel, and I started to pass him our gear. "Here's the floor." I handed him a yellow tarp, eight feet on a side.

"In place," Dave said after he spread it out on the snow. I passed him our sleeping duffel kits, each with one-inch Ensolite ground pads, sheepskins, and sleeping bags. We'd sleep just a couple of inches above the stiff snow.

Next, I passed him the radio, then the wooden boxes of food and cook pots. The radio went up by our heads. The boxes he arranged down the center of the tent between our sleeping bags, with a shelf between them to place a one-burner Optimus stove. I ducked my head into the tunnel and watched him, decided he had this handled, and let myself relax.

While he did that, I piled snow on the tent skirt that lay on top of the snow. The skirt had to be buried deep enough, a few feet, to keep high winds from getting under the tent bottom and inside. If that happened, the tent would blow away like a real umbrella, more like a parachute. That had happened to other teams.

"Ready for the donkey dick?" I asked. Dave exited the tent and grabbed a wooden supply box for me to stand on. A gray rubber hose, about four inches in diameter and a foot or more long, needed to be slipped into a sleeve at the top of the tent to act as a chimney. Without a chimney, carbon monoxide from the small stove would collect inside the tent. That could mean big trouble, like death. Fumes from his cook stove almost killed Richard Byrd during his sojourn alone on the Ross Ice Shelf. The hose would make a peephole to look outside when we tied shut the tent entrance tunnel in a storm. I stood on the box, and Dave grabbed my jacket while I leaned over to the peak of the tent, grabbed the opening sleeve, and inserted the tube. We stood back and admired our finished product. Home sweet home. We grinned at each other.

I noticed a pain in my right groin as I finished. Damn, that hurt— not again. That pain had appeared first in Santa Barbara. Now it returned at an awkward moment. I sat down to rest. Dave crawled out, and I told him I might have pulled a tendon in my groin. He gave me a look to say, that's bad timing. He was right. I needed to handle this, see a doc in Mac Town. I couldn't be laid up in the mountains.

Behind our camp on a low, black outcrop, a tall wooden cross cast a shadow to the north. At first, I assumed this cross was raised as a memorial to Scott's party of five men, him included, who froze or starved to death on their way back from the Pole in 1912. Bodies not recovered. No, that cross stood on Observation Hill—a small, dark brown volcanic cinder cone behind McMurdo Station. I'd hiked up it a couple of days ago and read the inscription from Tennyson. The cross here, at Cape Evans, memorialized three men of Shackleton's Ross Sea party, members of his 1915–1917 *Endurance* expedition who stayed in Scott's hut. They died while they waited for Shackleton to cross Antarctica from the Weddell Sea in the southernmost Atlantic. Bodies not recovered. Shackleton never crossed Antarctica.

Unknown to them, Shackleton's ship, *Endurance,* and crew had been trapped in sea ice and marooned for over a year. I had seen another cross at Hut Point in McMurdo next to Scott's 1901-1904 *Discovery* hut. That cross remembered a man lost on that earliest expedition, first to die in Antarctica. Body not recovered. I figured nine men, maybe Christians who needed crosses, had been killed here, and that earned them three crosses on three hills in the last place on Earth.

Tents set up, we walked over the snow and down to the *Terra Nova* hut. I'd read about it, built by Scott in 1911, occupied until 1913, then by Shackleton's Ross Sea *Endurance* party. Snow had drifted against parts of the uphill side, but other walls were exposed. The hut had been buried in snow since Shackleton returned here in 1917 to rescue the remainder of his Ross Sea party, who had been marooned in the hut. Americans dug it out in 1956. A Kiwi group had been restoring the hut since then. I knew we soon would stand next to history.

The hut looked to be in good condition. No free water, bacteria, or fungus to cause deterioration, and low Antarctic humidity preserved it. All walls stood intact. Outside, empty wooden supply crates were laid up against the sides. A skeleton of a dog spread out near one wall. Scott's men had built a low, closed-off covered shed attached to one side of the hut. I knew the doomed ponies they'd brought had lived and died there.

We circled the hut. I imagined the hard experiences of those who had lived here almost eighty years before. The hut door had a padlock— we couldn't enter, but we could see in through a small window on the uphill side. I crouched, got on my knees, and peered inside.

Things appeared in perfect order. Had they just stepped out? In the dim light, I made out bunks and clothing hung on walls, cans of food on shelves. A long table stretched out in the center of the main room. I recognized this table as the centerpiece in a photo of Scott and his team celebrating his birthday some months before they would leave for the Pole. He had sat at the head with a banner of flags hung on a line behind him, his Last Supper. Brave or foolish? Did that matter?

The wooden walls had black shiny tar on them in many places, smoke residue from burning seal blubber for fuel. The room looked small, of course, but had held over twenty men of Scott's expedition for almost three years of cold hell—the first space station. This proved to me that humans could make the trip to Mars. Incredulity set in, and I shook my head. That would be easier.

We were silent, did not share what thoughts we had. Did these men see themselves as heroic? I doubted that. Probably just like me, they sought new knowledge and a life experience with no equal. Whatever their motives, for a moment I felt trivial.

With no important words to share, we walked in quiet up the snowy hill to our tents, food, our sleeping bags, and rest.

CHAPTER 7

Shackleton's Room

Men Wanted for hazardous journey. Small wages, bitter cold, long months of complete darkness, constant danger, safe return doubtful. Honor and recognition in case of success.

—Attrib. **ERNEST SHACKLETON**, Recruiting Advertisement

San Diego, California, 1962

I waited in the damp shop warehouse at the Scripps Institution of Oceanography for the first meeting with George, my supervisor. My eyes scanned the cavernous space that held the aroma of the ocean and lubricating oil, looking for someone who I imagined as my boss. I shifted my weight from foot to foot. A man headed my way at a good pace, weaving around worktables. Huge, his muscles bulged from a dark sweatshirt, wavy blond hair, clean-shaven; his face looked sunburned and a bit flattened. The man named George looked down at me.

"So, you're the college boy. Huh," George said with a stern face. "How old are you?"

"Nineteen." I wondered if I should say sir.

"Nineteen. A fucking recruit."

I froze, startled at this unwelcome news. He cracked a half smile. I laughed, weak.

"Follow me," he said, then turned around. He didn't care if I was going to college. Until I proved myself, I'd be a recruit.

I'd taken a six-month leave from college in 1962 to go on a research voyage with Scripps to the western and southern Pacific. Before the expedition sailed, I worked at Scripps, assigned to help George. George taught me—about accountability, and about Shackleton. For that, I admired him. I wanted to hang around him. Retired navy, maybe fifty-ish, he had a temper and a sense of humor; he shouted with emotion rather than spoke much of the time.

Scripps was in the diamond-encrusted seaside village of La Jolla, across Soledad Mountain from the grinding, busy navy port city of San Diego. George and I got together for brown bag lunches a few times, sat at a redwood picnic table on grass atop the low cliff that overlooked the Scripps beach. We could see and hear the surf, smell the salt of the glorious Pacific. Two other marine technicians joined us most of the time. Fred always chewed on a cigar. He stuck it straight out of his face tilted down, parallel to his large, sharp nose. Fred looked older than George, heavy, with a large belly, a dark weathered face, and white hair. Fred had a taciturn demeanor that contrasted with George's volatility. He decided to call me Junior or Sonny, not Bruce. George picked up and used Junior except when I made him angry; then he called me a "fucking recruit." The other tech, Mack—thin, red hair, forty-something—had lied about his age to get into the navy at fifteen. He showed me his tattoos: a chicken on one foot, a pig on the other.

One lunch, George mentioned a book he'd read.

"Just finished *Hunter*, about big-game hunting in Africa early this century."

"Oh yeah, great book. I read it last year," I said.

"Really? Surprised a little punk like you would read that." I hid my smile at his unsurprising remark.

"I'm reading Shackleton's book now," he added, appearing to test my knowledge.

"Who's Shackleton?" I asked. George turned his head sideways to look at me.

"Aren't you goin' to fuckin' college, doncha know Shackleton?" he shouted. Fred and Mack sat across from us. Fred blew smoke. Mack snickered and smirked. They knew about Shackleton.

"Tell me," I said, a wise response.

"Read the fuckin' book, Junior," he said in a normal voice. "His ship was crushed in the ice near the Antarctic Peninsula, in nineteen-fifteen. The crew got marooned for months. He knew they'd all die unless he sailed a lifeboat to South Georgia, across the Drake, got help." I did quick geography in my head. He meant the Drake Passage between the Peninsula and South America. I'd heard about it, the most dangerous stretch of water in the world's oceans. South Georgia had to be a thousand miles from the Peninsula.

"They made it, got a ship, went back, rescued his crew. Nobody died, after over a year," he finished, impatient with me.

"What? Really? Are you shittin' me?" I said. The story sounded so crazy, so hard to believe.

"Don't be a fuckin' recruit," George said, soft, didn't shout this time, looked away.

Fred pulled his cigar out of his mouth, let out a puff of blue that drifted in the clean breeze. He looked past me, said, "Read the book, Sonny. You'll learn what the word courage means."

Santa Barbara, California, 1984

In 1984, I discovered Shackleton's book, *South*, by accident while browsing the local public library near UC Santa Barbara where I taught geology. I realized I had found the book George mentioned twenty years before. It had photos in a center section. A tall, rigged sailing ship, the *Endurance*, trapped in the ice of Antarctica's Weddell Sea, then splintered and crushed, leaving men and tents on the ice, lifeboats too.

South told the story firsthand. After moving ice crushed their ship, the *Endurance* crew of twenty-eight men traveled north in lifeboats to Elephant Island near the tip of the Antarctic Peninsula. In late April 1916, Shackleton and five of his crew sailed northeast some nine hundred miles in a small lifeboat to South Georgia Island, made it to the whaling station there, and began the long task of a successful rescue in August 1916.

I closed *South*. I had just read a story about courage; it happened in Antarctica. I wouldn't have had the nerve to go with Shackleton on his adventures. I also knew that if I had been in his shoes, I, too, would have stopped at nothing to rescue my people. No uncertainty at all about that.

Shackleton wanted to achieve an exploration goal: to cross Antarctica. In a few years, I would lead a geological expedition to the wilderness of Marie Byrd Land. Still, I imagined the depth of scientific unknowns about such an alien place, a place that drew me like my urge to explore the unseen ocean floor. I could create knowledge about Antarctica, a barely known land.

But I also learned this from Shackleton:

I must go to Antarctica. I must be courageous.

Cape Royds, McMurdo Sound, Antarctica, November 24, 1989

In the hard morning work of breaking camp at Cape Evans, I had strained my groin tendon once again. I'd complained to Dave.

"That might be a problem later on, Bruce," Dave said, and continued rolling his sleeping bag into a tight ball, using his substantial body size to his advantage. "We're just getting started, with a couple of months to go. Better get on top of that." He usually didn't give advice.

He was right, but that had to wait. Chris had planned our present crucial shakedown trip for our expedition to Marie Byrd Land with visits to historic explorer huts on Ross Island. She didn't know visiting Shackleton's *Nimrod* hut today would fulfill a dream I'd nurtured for twenty-seven years, since I was nineteen years old.

Our group of seven had left Cape Evans and Robert Scott's 1911 *Terra Nova* hut near noon. After our first experience camped out overnight, we now loaded up our gear and tents on snowmobiles and sledges and got underway to Cape Royds.

Shackleton's Hut at Cape Royds 1989.

We drove over the sea ice under a layer of gray clouds. To our right—east—the enormous cone-shaped white mass of Mount Erebus spewed volcanic steam from its summit. Breaks in the clouds allowed speckles of sunlight to illuminate the gray ice. They moved with the clouds in the breeze. We passed a few black volcanic islands that poked up through the ice, reaching heights of hundreds of feet. Even with the overcast, the sterile air allowed us to view the majestic Transantarctic Mountains more than twenty miles to the west. The scene told me to remember this special adventure. Forever.

My groin hurt as I bumped over windrows of snow scattered on top of level, slate-colored ice. I consoled myself: *I'm going to see Shackleton's hut.* A cruel southerly breeze had blown at our backs. The weather had not changed much from yesterday: cold. Even wearing all my ECW, I crouched low behind the windscreen on my snowmobile. My nose went numb, and I tasted salt ice on my beard. A couple of hours and seven miles farther north, our train of four snowmobiles, each pulling sledges, came to Cape Royds, a low promontory of dark volcanic rock shaped into rolling hills.

We stopped on the sea ice where it met the edge of land and prepared to visit Ernest Shackleton's hut. I struggled off my snowmobile, unbending stiff limbs. Taking off my bear-paw mittens, I shook nearly numbed-out fingers. *You're here, Bruce.* I ignored my tendon pain and fingers and prepared to live my dream.

I watched my footing as we picked our way up a low slope over black volcanic boulders and gravel to a saddle in a ridge. From there, the hut, a penguin colony, and a frozen lake came into view. Behind these, hills of glacial ice framed the scene. The *Nimrod* hut used during Shackleton's 1907–1909 expedition, before his 1914–1917 *Endurance* expedition on the other side of Antarctica, stood about two football fields away. Between me and Shackleton's hut were penguins, hundreds and hundreds of penguins. In front of me and a bit to the right, Earth's southernmost colony of Adélie penguins spread out on several low black-and-brown knolls like so many large ants. They squawked, squeaked, and stunk like old fish. We followed a thin, brown dirt pathway down the slope before us to the frozen lake at the edge of the colony.

Hard to ignore the penguins, so many of them; they squawked and waddled about, making a comical, if classic, scene. We walked up to a nest area, but not too close. US law states if you cause any Antarctic animal to move in avoidance, you break the law, a felony. I spotted a few short stakes with colored markers in the ground. Biologists had worked here.

Every penguin waddle made me smile, like infants that coo and squirm. Amid nests made from volcanic pebbles, penguin mates stood near each other. Brown-pink guano covered everything. We placed our steps with care. Clean birds hopped about; they had been washed from fishing somewhere nearby, through holes in the sea ice—holes we didn't see on our drive but missed. Lucky us. Other birds had soil and guano smeared on their feathers. Done with cuteness, we watched their behavior; they challenged their neighbors, protected their mates. Some sat in nests, maybe on hidden eggs. I didn't see any chicks, too soon in the season. None of the penguins acted curious or frightened. I took photos.

We ended our close encounter and walked toward the hut. The *Nimrod* hut looked smaller than the *Terra Nova* hut, about thirty feet long or nearly half the size—this hut built in 1907, before the other at Cape Evans. On his *Nimrod* expedition, Shackleton and three men struck out from here to discover the South Pole, three years ahead of Scott and Amundsen. They made it to within one hundred miles of the Pole and claimed Farthest South. Shackleton decided to turn back when he realized they would starve and die if they went all the way. "Better a live donkey than a dead lion," he later told his wife. I thought that remark very heroic. I found myself mouthing it while I surveyed the hut as I approached.

Scott's earlier *Discovery* expedition in 1901–1904 had attempted the Pole but also fell short. Shackleton accompanied Scott on that first unsuccessful attempt. Shackleton got scurvy, ending the quest. Scott publicly humiliated Shackleton. He blamed his illness for the failure. I wondered what it felt like to Shackleton to be criticized for risking his life for another person's glory. Later, on *Nimrod*, Shackleton went farther than Scott and got knighted for it. I had mixed feelings about Scott. He was heroic but proved to have made rash and deadly judgments. Shackleton, to me, seemed strategic. I could identify with his approach.

Shackleton's Ross Sea party waited for him to cross Antarctica from the Atlantic side—the *Endurance* expedition. Ten men had waited in the *Terra Nova* hut and at the *Discovery* hut south at Hut Point, but Shackleton didn't come as planned. After rescuing his men on Elephant Island, Shackleton participated in their rescue and reached them in January 1917, but three of them had died, their cross now on the hill behind the *Terra Nova* hut. I reflected on the impossible uncertainty and loneliness those men suffered. This part of Shackleton's famous adventure is not well recognized or appreciated. As I stood there, knowing this gave me a lump in my throat.

I snapped a photo of the hut from my vantage point of thirty yards away and a little above it, then stuffed my camera in my pack. Antarctica had a way of making a person feel tiny, invisible—a wave of insignificance swept over me. Shackleton's towering accomplishments were out of scale with this hut that kept him alive, where he must have had moments of knowing he could die like other explorers who tried and failed. As I began my first expedition in Antarctica, I did not want to try and fail—not for lack of courage.

Similar in design to the *Terra Nova* hut, the *Nimrod* hut had supply crates stacked along the outside walls. Two steel cables crisscrossed the top of the hut and anchored it to the ground. Shackleton had remembered to prepare for the inevitable ferocious storms. We walked closer toward the north side along a collapsed frame of stables and bales of hay. Rows of boxes outlined the vanished stable walls. We approached the west side that faced the frozen lake and sea ice beyond, mounted a few steps to the front door—padlocked.

"I have the key," Chris said. She came forward, unlatched the padlock, swung open the front door.

"Oh, wow," I said, in a whisper of surprise only I could hear. I hadn't expected we could enter.

She stepped inside, I followed, then the rest of my team and Jill. We stood in a covered porch, the vestibule and weather lock. Another door faced us, not locked. Chris pulled it open, crossed the threshold, and all of us entered through a short dark hallway. We gathered in a small group at the near end of a large room. We didn't speak. What could we say in a moment like this?

An open space appeared before us. A good light came in through two north windows. We took a few steps. Our rigid mountaineering boots clunked on the wooden floor, made soft echoes. Twenty-one years later, in 2010, restoration crews would find crates of whisky and brandy under these floorboards. We stopped, looked about. No spiders or cobwebs. More than a dozen men had lived in here.

"We can't touch anything," I said. Nobody acknowledged my quiet statement.

Metal cots and wooden beds stretched along the hut sides; an iron stove sat at the far end. Nansen sledges hung from the rafters; skis hung along one wall. Ragged, decayed bedding covered cots and bunks. Boots stood upright on the floor and hung on the wall with coats and socks. One curtain separated bunks. Stacked against walls were wooden food boxes labeled "British Antarctic Expedition 1907," the official name for

Shackleton's Room

Nimrod. Shelves held cans of beef loaf and jars of jam. Crates of biscuits and other food lay scattered on the floor everywhere, packaging still intact. Cooking gear, pots, and a tea kettle remained set on the stove.

Pictures of King Edward and Queen Alexandra stared back at us. Giants once lived here: Edgeworth David, Douglas Mawson, Raymond Priestley, and Ernest Shackleton. I recalled his legendary recruitment advertisement: "Men Wanted for hazardous journey…" *So, this is what courage looks like.* I turned to my right; a walled-off cubbyhole with a narrow doorway in the southwest corner framed a private space. Shackleton's room. I hesitated, then approached, looked into a dark volume. Eyes adjusted, when I recognized a bunk and table, my scalp crawled. My breathing stopped; my torso felt electrified. *He lived in here, the man who led me to Antarctica.* Almost ashamed by my trespass, I got lightheaded and needed to leave, to not linger in that sacred space. I thought about what was expected of me, how and what I would need to deliver in just a short time. Still, I couldn't compare anything that faced me with the challenges that Shackleton and his men accepted gladly.

Chris and Dave came up behind me, stared into the dark space for a moment, then moved on. Overwhelmed, I backed away, found the hut entrance, and withdrew outside, my chest in knots. The rest of the group followed me out. I glanced at their faces and could not read their feelings. All of us knew about Shackleton, but maybe I was alone in my reaction.

We walked in silence up a low black hill behind the hut, down the other side to the sea ice, to our snowmobiles and sledges. I checked my camera photo count. Same number as after my visit to the penguins. I hadn't taken pictures inside Shackleton's hut. I didn't need any. I hoped I'd never need to face the scale of the challenges that Shackleton did. What did the rest of the team feel when they gazed into the dimness of Shackleton's room? I studied them as we prepared to saddle up. No clues, no remarks. I took a giant breath, filling my lungs with the pure, frozen afternoon air. Glad to be alive. My snowmobile roared to life with my first powerful pull on the starter cord. I was ready for whatever lay ahead.

CHAPTER 8

Debrief

You wait. Everyone has an Antarctic.
—THOMAS PYNCHON, *V.*

McMurdo Station, November 25, 1989

The thermometer clipped to my Big Red zipper sat at minus fifteen Celsius. The weather turned bad when we left Cape Royds. I put on all the clothes I brought with me. After a stop at Cape Evans, our team, with Jill again riding prone atop a sledge, continued south over the sea ice headed to McMurdo. We drove into a headwind, the sky gray, clouds broken in places, blue sky patches; dapples of light swept across the ice, fast with the wind. Barnes Glacier by Cape Evans glowed blue when a piece of the sun shone on it. Jet-black volcanic crags of the Razorback Islands jutted straight out of the gray-blue-white sea ice; their serrated crests cut into the sky. The warmth of the beauty battled the grating cold.

Wind came from the Polar Plateau down through glacier passes and then flowed north onto the sea ice and ice shelf, into our faces. Frigid, but the wind could've been worse, a fierce katabatic wind of hurricane force, or a Herbie, the nickname. A bitter cold mass of air on the Plateau gets too heavy to hang out up there and collapses over the edge, rushing down the slope like a formation of jet fighters. Better get shelter quick if you are caught out in one—the windchill would be deadly. But there was no shelter near us on the sea ice.

My outfit: two pairs of socks, my Bunny Boots, red wind pants over fleece and long johns, Big Red parka over wind jacket, fleece jacket and long johns and undershirt, balaclava on head, face mask, goggles, parka hood pulled tight, shaped by wire inserts within its border into a fur-lined tunnel. On my hands, ski gloves over glove liners that were then stuck into clumsy, oversize fur-and-denim bear-paw mittens.

Hard to recognize one from another dressed for protection like this—we needed to be close enough to see those white Velcro strip nametags on the Big Reds, or pick out distinctive body language, like Dave's massive size and shuffling gait, Chris' bounce, Cain's rocking stride. I tried to make myself small, crouched behind the windscreen on the snowmobile, but it didn't help. Pain assaulted my face and fingers. I watched Jill as she lay flat on the sledge ahead of me, arms and legs spread to grab on, watching for a sign of trouble. She didn't complain. *She might wait too long to do that. Then what?*

Cain took the lead. I believed he drove too fast, and I couldn't get his attention. When he stopped the group for a break, I told him how I felt, and this time we were driving into a headwind, more frigid than the downwind trip a couple of days ago. I asked him why he believed it necessary. His response sounded dismissive and vague. He turned away. He ignored me. We got back underway. Cain led at the same speed.

A few hours later, we arrived at the Transition in McMurdo. Jill rolled off the sledge, stiff. I watched her, but she appeared to walk okay. We began the process to unload—brutal cold, that trip. My arms and legs felt leaden, hard to move at first. Would the cold be worse in the mountains? It might be warmer near the coast, but we were on the coast now. Regardless, I realized that I had just done special things. I slept in a tent in Antarctica and visited huts of legendary explorers. I granted myself some gratitude.

I stood next to my machine on the ice and motioned for us to gather around. "We need to debrief ourselves on this trip, find out what to fix, to improve," I said. Murmurs of assent, but all of us were too beat-up now to think about those topics. "Tomorrow morning in the dorm?" I said. "After chow, eight a.m.?" All agreed.

Later that evening, I relaxed in my comfortable dorm room. My groin throbbed. I decided I needed to go to sick call soon and get that looked at. I could end up a burden on the team.

I stretched out on my bunk, alone; Dave disappeared elsewhere. *This room is heaven.* I had discovered what we'd be up against, both the wonder and the pain. I got out my journal, my no-freeze pen, and made two columns: Good, Bad.

Good: constant daylight—but need eyeshades to sleep, no insects, snakes, or bears, no dust, plenty of water once you melt the snow, scenery to die for, real wilderness, good companions. Bad: it's cold (I looked at that for a moment, crossed it out), cold as hell (still didn't look right—crossed that out), it's fucking cold; bitter winds, tents are cramped, noisy in the wind—earplugs required, no place to sit, take a dump outside in the snow, hard work to cook and clean up, hard work to make and break camp, hard work to travel, no shower, no bath, can't do laundry. I read this over; it was a breakeven list. I snorted a laugh. I'd revisit the details later after some time in the mountains.

* * *

We met in the dorm lounge after breakfast, a corner room on the ground floor with a view of frozen McMurdo Sound through leaky windows that whistled in the wind. Venetian blinds broken or bent told a tale of parties and heavy frustration. Monday morning, a tall trash can full of beer bottles and cans sat conspicuously in a corner. Empties lay scattered across tabletops too. A pool table in the middle had a rip in the green felt, patched with duct tape. The team gathered around in stuffed chairs and a brown cloth sofa with a population of rips and stains.

Breakfast had been busy, but I detected tension. When we sat down as a group, I felt an active silence. Something was brewing. Our team had just completed our first trip in conditions that were a challenge but manageable. Did the experience compare to what we might face in the remote wilderness of Marie Byrd Land? The rough spots needed to be fixed.

I broke the silence. "What's on people's minds? What's missing, needs to be changed, fixed?" Lots of input came up. Chris, Tucker, and Cain had the most to say. We covered many topics, no surprises—we needed repairs and modifications to the snowmobiles, studs on their treads for ice travel, rope brakes for the sledges, a system for loading the sledges, and canvas covers, or tanks, for the sledge cargo. Why not customize the tents? I took notes; others volunteered to get these things done if possible.

Soon, silence took hold again. I could tell things went unsaid. I prodded. "What else?"

Tucker spoke up, said he felt frustrated, that we needed to communicate better. It didn't seem to him that we discussed what we wanted to do and when, too ad hoc.

"Why don't we commit to meet every morning before gettin' underway and set goals, like a tailgate party," he said—spoken from his experience as a contractor. A great idea, but I didn't hear a clear reaction from the group. Why not?

Chris and Steve chimed in that they wanted to be sure the Fosdick Mountains metamorphic rocks and their high temperature and pressure history were our priority. They didn't feel we should spend much time elsewhere studying igneous rocks, those that crystallized at depth, mostly undisturbed, and bystanders to the turmoil of the metamorphism of the Fosdick rocks. Now I got some insight, a conflict of goals; they had their own plan and wouldn't need tailgate meetings. I didn't react to that news, looked at Dave for some direction and couldn't detect any.

Cain brought up that he heard too much complaining. I felt sure he meant me. I took that bait. "Here's a complaint, I think you drove us too fast over the ice. Why was that? It didn't seem necessary. That wasn't safe. Jill had a hard time hanging on. What if she fell off?" The team turned their heads, looked at me. I realized right then I had crossed the line. I said he acted unsafe. That was not wise of me; I went too far. I couldn't judge what was safe. I didn't know. An accusation of acting unsafe to an Antarctic guide from a PI, me, could cost him his reputation, job, career. What would happen now? My forehead got damp.

"I didn't hear her complain, just you," Cain said, clear through his Scottish brogue. He gave me that look of his, disdain, which I had learned meant: you are out of your depth here, Bruce; get a desk job if you don't like it.

What do I do? He just acted insubordinate. I had been in situations like this on marine expeditions. On paper, the ship's crew worked for the science party and Cain worked for me, but they had power because they knew how to do stuff, had the experience, and I didn't. The group remained silent, waited for the next salvo. I cleared my throat to speak, but then decided to defer, not respond.

Tucker spoke up, said this exchange should show us why we needed better daily communication. We had a lot to do—Survival School was in two days, so let's get busy. He looked sideways at me, raised one eyebrow. I got his message: what the hell do you think you're doing?

Looking back over the trip, I realized I could have cut Cain a break. He was the expert after all. I was the greenhorn. My irritation at him might have sprung from the pressure I felt having responsibility for the project, and even for the safety of everyone. Fair enough, but damage control needed to be on top of my to-do list. I rubbed my face with both hands. I would need to find a chance to apologize to Cain, then make a team out of six people with different agendas.

CHAPTER 9

Survival School

No matter where you go...there you are.

—**BUCKAROO BANZAI, Ph.D,**
Neurosurgeon, Theoretical & Practical Physicist,
Race Car Driver, Rock & Roll Star, Comic Book Hero

McMurdo Station, November 25–December 3, 1989

"I heard they loan out videos from a library somewhere around here, and there's a player in the dorm lounge," I said. Our team sat at a table in the Galley. I tried to think of a way to bond after the debrief confrontation. "Want to watch one before we head out to the bars?" Four bars offered a way to pass time in McMurdo. Tucker said the video library was behind the laundry, and he promised to find a good one for us.

In the dorm lounge, a dusty TV and video player sat in a corner. A few people were reading; we asked if we could show a video. Permission granted. We did our best to dim the windows with the broken blinds. I dusted the volcanic grit off the TV screen, rubbed the fine sand between my fingers. Next to the TV, I saw a video, *The Last Place on Earth*, which I recognized as the BBC production about the Amundsen-Scott race for the South Pole. Amundsen had beaten Scott by several weeks. I announced my find, recommended it, and got some interest from our team and the few folks in the room. Tucker came in as we started to watch it.

Everyone paid rapt attention. The film depicted life in polar conditions with great accuracy. Also accurate were the scenes of death and the frozen bodies of Scott and two companions that were found in their tent a year after they died—from starvation and scurvy, eleven miles from their next supply depot. The pain of the deaths, the discovery of the men, the burial by their teammates of the tent with the bodies under a pile of snow—all were vivid.

In the last scene, Amundsen is in his bathtub at his Norway home a year after his Pole victory. His brother Theo interrupts him, reads him a telegram. The bodies of Scott and companions had been found. Amundsen stares into the bathwater and says, "So he won the race after all." I had forgotten the grim nature of the film and felt a bit unnerved. *Of course, we won't be up against situations like theirs. Will we?*

Video ended; the room became quiet. "Thanks a lot for the bummer, Bruce," someone said. "Yeah, now I know what freezing to death would be like," said someone else. I had made another blunder. I remained silent.

Tucker jumped up and told us he had a video to change the mood, said laughter is good medicine. I waited for his surprise. He told us the title as he crouched down and inserted the tape, *The Adventures of Buckaroo Banzai Across the 8th Dimension*, a sci-fi comedy he assured us. "Never heard of it," I said. The rest of the folks looked in the dark too. It started to play and got weirder with every minute.

Dr. Banzai, a physicist, adventurer, and Renaissance man, is tasked to save Earth from an alien invasion of the Red Lectroids and their archenemies, the Black Lectroids, both from the eighth dimension. All the aliens take on human life-forms with first names John. Dr. Banzai has an array of sci-fi machines at his disposal and can drive his modified Ford F-350 pickup straight into and through a mountain without harm. Turns out the Red Lectroids had been on Earth for a few decades, having slipped in during the radio broadcast of Orson Welles' *War of the Worlds*.

Welles is alleged to have known that but covered up the truth, claimed his broadcast dramatized H. G. Wells' novel. In the meantime, all the Red Lectroids had been working at a US defense factory. The Black Lectroids were disguised as Rastafarians. A Red Lectroid underling played by Christopher Lloyd is John Bigbooté. He corrects the pronunciation of his name in every scene. "It's John Big booTAY," he explains. Annoyed at this endless correction, his Lectroid boss shoots him.

The movie proved impossible to follow, but every scene had something crazy to set it apart. The script called for a recurring line: "No matter where you go, there you are." Tape ended, Tucker jumped up and said *that's our mantra.* "What is?" several of us asked. The team shared laughter. That was good.

"No matter where we go, there we are!" Tucker said.

"But what does it mean?" I asked.

"Whatever, of course!" Huh? That made no sense, but people chuckled, so mission accomplished, mood lifted. Tucker would continue to use this line throughout our expedition when things turned to shit or were about to. Then I knew what he meant.

* * *

Survival School started the next day, Monday. We needed three days of school because we were designated a Deep Field party and would travel over glaciers in our work. These days, the course is named Snowcraft, and the Mac Town slang name for it is Happy Camper School. Why the name change? My guess—remove the edge of intimidation from the experience.

Day one arrived with snow and hard wind, temperature near freezing. The snow began to melt. I stood at the dorm door, looked out. Conditions were miserable for sure. *Maybe they'll cancel?* Nope. My body felt heavy as wet snow. Tucker, Steve, Dave, and I loaded into a Delta, a brutish, pale orange truck with a steel box on its bed for carrying passengers on benches along each side of it. Its tires stood head high. Cain and Chris had done school the week before the rest of us arrived. Today they worked on our Put-In supplies. The vehicle started the jarring ride out to the school site past the New Zealand Scott Base. We lurched onto the sea ice, then bucked and rocked up to the front of an ice slope of Mount Erebus.

Our personal rucksacks were packed with climbing gear. Looked like neat stuff; I hoped I wouldn't have to learn how to use or need all of it—wrong. Sixty meters of climbing rope, a climbing harness belt with thigh loops, carabiners (clip and locking types), pulleys, ice screw anchors, mechanical rope ascenders (jumars), Prusik rope ascenders, snow anchors, crampons fit to personal boots, and ice axes.

I felt cool suiting up and carrying all this gear. My mood brightened despite the snowy gloom. Made me think I looked like a mountaineer, but I didn't have any significant experience. I hadn't seen snow and ice gear before.

We arrived at the school site and joined a small group of young, navy enlisted men, our classmates. They stood around in the snow, joked and smoked. The smell of tobacco overlay the scent of driven snow. These guys had bulky navy-issue clothing and huge, white rubber Bunny boots. They were olive drab Michelin men, wore no classy mountain clothing or boots like ours. A couple of instructors greeted us, appeared to me like they had just descended Everest.

"Welcome. We'll start with learning how to stop a fall on a slope, without ice axes," one said over the wind and snow that blew sideways. Wet icy globs pelted my cheeks. *I should have called in sick.* "Idea is to get on your hands and knees, head uphill, from an uncontrolled slide down a slope, and stop," he said. "Yeah. You guys with the fancy clothes might slide faster. That's smooth synthetic fabric you got on I see." He pointed our way. I wore a blue wind parka and red wind pants, both a bit shiny. A couple of navy guys smirked at us, puffed their smokes. They didn't respect Beakers.

Our group of students slogged about a hundred feet up a snow slope. The instructor showed us how to stop after a fall on your back, face, butt, head downhill, head uphill, and so on. This felt like play to me, except for the weather. First try, I fell on my back with head uphill, flipped over to hands and knees, caught the toes of my boots in the snow, and went head over heels onto my back, then tumbled down the slope to the bottom. I made a snowman of myself. I could hear some guffaws from those at the top, navy guys I thought. I ignored them. We practiced for a couple of hours. All twenty of us got covered in snow. I got soaked from it melting inside my clothes and started to shiver.

Lunch break took place in blowing snow: trail mix, chocolate bar, hard salami, and frozen cheese slices. Then we got out our ice axes. "You'll be walking on slopes with an axe, so the idea is to use it to stop a fall and not spear yourself with it; learn to self-arrest," we were told. My ice axe measured three feet long, violet with a silver barber-pole stripe. It looked dangerous. I liked that.

"The axe is designed for different jobs," our instructor said. "One is to save your life." The instructor picked up his axe and held it up for us to see. "This is the head. It has a chisel blade on one end, the adze, to chop footsteps. The other long pointed opposite end is the pick, with a sawtooth lower edge." He ran his hand over the head and parts. "The pick is what you use to self-arrest." That end looked like it could spear me straight through.

He held up the bottom of the shaft. "This metal point is the spike, at the end of the shaft. Use it to probe for crevasses and to stabilize yourself when you walk on a slope." He planted the axe into the snow. "Put your downhill hand through this wrist leash attached at the top of the shaft. If you fall, the axe will go downhill with you. So, it's important to use it for your safety and not against."

The instructors demonstrated adroit maneuvers with an axe while sliding downhill. Worst-case fall is on your back with your head downhill. "To self-arrest, plant your axe pick to spin you around to head uphill, flip over onto your belly, to lay atop the axe, then dig in the pick and stop," he told us. "Go to it."

We dove and jumped down the hill in varied positions, used the pointed end of the axe head to dig into the snow with our body on top of it. Looked like we all got the hang of this—didn't stab ourselves or poke out an eye. *If I know I'm going to fall, I can do this. Not so if I fall unexpected and panicked.* Tucker had done this stuff before. He offered encouragement. "You guys are gettin' it," he said in a calm, measured voice. I showed Tucker a thumbs-up. He nodded back.

The instructors gathered us around late afternoon. "Okay, now you're goin' to build a snow shelter to spend the night here," he said. "Two choices we recommend, snow trench or snow mound." He surveyed us to see if we paid attention. Snow continued to fall. I cleared my throat a couple of times and sniffled the wet snow. I didn't expect an overnight would be enjoyable.

"Ambitious guys can build an igloo, but that'll take about four hours," the other instructor said.

The four of us decided to build a snow-mound shelter, called a quinzhee; it sounded quick, easy, bombproof. Make a six-foot-high pile of all your gear and bags, bury these in about two feet of snow, pack it down hard, tunnel underneath from outside, and pull out all your gear to leave a cavity to sleep in.

Christine had told me that she and some random guy in her Survival School class the week before had built a two-person snow trench, an eight-foot-long slit a few feet wide, five feet deep, with shelves for beds carved into each wall of the trench. They cut snow blocks to place over the trench and make a roof.

Done and ready for the night, her new companion turned to her and said, "Betcha haven't had sex in a snow trench before. Wanna go for it?"

"No, I haven't, and no, I don't," she had replied, shocked. Surprised, he persisted, pointed out this chance for a unique experience. He meant bragging rights for him. She made up a story that she had a boyfriend.

"Who's gonna know or tell?" he said, then stripped naked, slipped into his bag near the wall opposite her. "If you change your mind, I'm right here."

She told him, "Don't make me get out. You'll have to explain that." She lay awake all night, one eye open, listening to him snore and fart. Hearing the story, I wondered why women are interested in men at all—so many of us are assholes.

Almost done with our shelter, we watched as the snow stopped, clouds opened, and a van drove up with Chris and Cain. "Weather cleared in Marie Byrd Land. Our Recce flight is on now. Pack up," Chris said. Great! A flight out to the mountains would be more fun than a night in a cold snow cave in damp clothes. In haste, we uncovered our gear, full of wet snow, threw it into the van, and headed off for the ice runway at McMurdo and a Herc ride.

* * *

The Recce flight took several hours. Most of the Ford Ranges lay under low clouds by the time we got there, so after a few passes we turned back. We had to schedule a repeat. We got to bed about four in the morning and then up at six for day two of Survival School. I felt pretty beat. For the second day, the four of us attended again without Chris or Cain. Driving out there, I asked Tucker, "What's up for today?" He wasn't sure, thought they'd ramp things up a bit, but not to worry, it would be straightforward—just needed concentration and coordination, so don't make a mistake. I knew mountaineers loved that aspect of their sport: to have no room for error.

We arrived and found the navy guys absent for today's class; it was us four and a few other Beakers. The instructors told us we would learn to use crampons and rope together in threes, travel across icy terrain, learn how to arrest a fall of the lead person on the rope, do a rope belay, then anchor the person who fell.

Three of us put crampons on our mountain boots and tied-in to a forty-five-meter rope, with one person in the middle. An instructor addressed us: "As you walk, keep the rope between you off the ice, but not so tight as to pull the others off balance. When I signal the first man, he'll fall down the slope, and you two drop to a sitting position on the ice and stop him with your feet dug in." We began. I tied-in last on the rope, we walked down the slope, and the front man fell. I dropped to the ice and got yanked forward over my crampons—one tore a hole in my pants, but not my leg. The other guy on the rope stopped the slide. "Do that again until you get it," the instructors told us, and we did.

After a few hours of practice, we heard, "Now on to crevasse detection and extracting yourself from one, if you're not hurt." Crevasses would prove the daily challenge in our fieldwork. I recalled the words of a mountaineer we visited in Christchurch a few weeks before. "Avoid slots. If you travel across crevasses, you're going to fall in and die," or something close to that. Crevasses crossed our travel paths and guarded access to rock outcrops that we needed to study and sample.

We marched over to a spot where a crevasse with a snow bridge had been located. My stomach knotted. "Stay roped, and the front man will probe with his axe to find the crevasse hidden under the bridge." We took turns at this. I walked onto the bridge and probed, found it measured about two feet thick. I looked over my shoulder to the other guys who held the rope I had tied into.

"Now we're lowering you into the crevasse over here," the instructor said, pointing to where the snow bridge had collapsed. "You two hold tight and walk forward and lower him into it." He meant lower me. They eased me down about thirty feet. I dangled in midair by the rope tied to my harness belt. I looked about in the dim light at the pale blue ice walls, the instructor's head outlined against the sky above me, the empty blackness below me. No sounds, just my breath, I rotated on the rope.

"Can you hear me?" I heard the instructor shout down. The snow absorbed much of his voice. I signaled yes but wanted to say, "Just barely."

"Clip your ascenders onto the rope and climb up," he said. We'd been shown how to do that. I clipped on my jumars and in a few minutes shimmied up the rope to the edge of the surface. Two men grabbed my armpits and hauled me out.

I felt exhilarated by my short adventure. I didn't get hurt or die in a crevasse. "Now, guys, realize that if he was unconscious or injured, you'd have to go down to get him," the instructor said. "That's for another class."

Third day, we united with Chris and Cain, and the six of us had the final class with a different field safety instructor. A young woman would teach us today. Pretty, maybe thirty, blonde hair and white smile, strong and cheerful. We put on crampons and walked to the top of an ice cliff about fifty feet high.

"We'll learn some snow anchors and pulley rescues from here," she said. Cain and Tucker would have to be patient with her. I was sure they knew as much as she did. "Tie in, and let's check each other's knots," she said. I asked Cain to show me once more how to secure my climbing belt and tie a figure-eight knot—the rope connection to the belt. I wanted a double-check for these exercises.

Cain handled the rope I needed to tie to. "First, make sure you've double-backed your belt loop." He showed me how, looked at my belt. I did it right. "To tie in, you can double the rope like a tie-in to a bight, or you can take one end of the rope and make a figure eight, then rethread it after passing through the carabiner on your harness," he said, then showed me how to do both. I copied what he did.

"That's it—good," he said, walked off to check the others.

The instructor came over to look at my setup.

"Hi," she said, grinned.

She slid her hand under my harness belt, pulled to see if I had threaded the loop right, handled my knot. "Looks like you know how to make a figure eight," she said, and smiled right at me. A nice encounter: I felt warmth sent my way.

The exercises included learning how to rappel over an ice cliff on a rope held by aluminum anchors hammered into the snow. "Can the anchors hold our weight?" I asked Cain off to the side.

"Windblown snow is hard. Look, your boots aren't denting the snow." He pointed down. I nodded and recalled hammering tent stakes into snow a few days before.

Confident that the snow anchors would hold, I rappelled down the cliff. Like the day before, we practiced with our ascenders to go up a rope and arrested falls while roped together. Class went smoothly. I deferred to Cain when I could, to let him know he held my respect in this arena. Tucker watched me. Maybe he realized my motives—show my trust in Cain.

The instructor paid attention to me, but I felt preoccupied. We chatted a few times as the day unfolded. I stood a moment for a break. She approached.

"Why don't we go for a hike tomorrow or the next day, you and I?" She showed me her broad smile. Caught off guard, I almost panicked. A hike with her probably meant up some cliffs and across crevasse fields. That would not be a date.

"Hey, that's a fun idea but we're goin' out in a couple of days and I won't have a chance," I said, surprising myself with how uninterested I was acting.

"Okay, I see," she said, detecting my awkwardness.

"When I get back in January maybe?" I said, trying to recover.

"Sure." She looked at me as if to say, I've heard that before. *Damn, what's wrong with me—she's cute and nice.* I remembered what my friend Mike had told me: "Bruce, I've noticed you're a bit clumsy around women," he'd said in his understated way with a proper English accent. Another mistake. Besides, I was single. But a new woman, Annie, had been on my mind. She and I were only a month into a fuzzy relationship before I left California, but I wanted to see where it would go when I got back. We hadn't agreed to be exclusive. Still, I didn't want to make my life more confusing.

Survival School ended; our team loaded up in a Delta to head back to McMurdo. This school made for good team building. Trust other guys on a rope for your safety—that matters. I had treated Cain with respect, Tucker with deference. This could work. The education experience of Survival School impressed me as a mixed bag. Glacier travel and rescue was complicated; we needed a lot of practice. I didn't see the opportunity for that. I learned that the others and I were still unprepared. We did have Cain and Tucker though. As long as they didn't go down and we had to rescue them, we'd be fine.

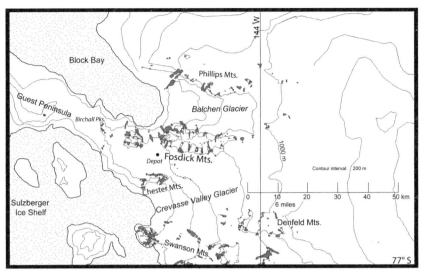

Map 4. *Mountains of the Ford Ranges mentioned in the story. Elevation contours are in meters (m). Ice shelves are thick floating ice fed by the inland glaciers. The main exploration activity occurred in the Chester, Fosdick, and Phillips Mountains. Depot was the main base camp for this story with the cache of supplies. (After US Geological Survey maps.)*

CHAPTER 10

Recce

In Antarctica, there's a woman behind every tree.

—ANTARCTIC SAYING

National Archives, Washington, D.C., Summer 1989

Months before the expedition, I calculated the distance from McMurdo to our fieldwork area in the Ford Ranges of MBL. It was almost the distance from Santa Barbara to Denver—eight hundred miles. My thoughts swam with the idea of nobody, no life, in the Ford Ranges, and nobody, no life, between McMurdo and there. I wanted to see photos taken by the first humans to look upon these mountains. I wanted to see what we'd be up against.

I slipped on a pair of lightweight white cotton gloves the clerk had given me and sat down at a large wooden table in a high-ceiling, wood-paneled room by myself—a viewing room in the US National Archives. It was the summer before we left and I had come to examine original air photos of our work area taken during the Byrd expedition that discovered this region in 1929. The clerk handed me a sturdy cardboard box. I opened it and touched a photo.

The photographs were black and white, eight-by-ten prints—they appeared fresh and not weathered with age. I realized I held originals.

These were history. My fingers trembled just a little. The scenes had cryptic lettering in white ink indicating dates of flights in 1929 and 1934. The pictures taken on these Byrd expeditions were the first record of these unknown lands. My concentration heightened as I inspected them. I daydreamed about the thrill Byrd's crew must have felt when they came upon these mountains.

Viewing the photos, I grasped the feeling of the remote and bleak aspect of the landscape and the challenges we would face. The scenes showed snow and ice, rock, and sky. A main glacier, the Balchen, named after Byrd's pilot, ran west from the ice cap of the interior to a floating ice shelf and separated two of the mountain ranges we planned to explore, the Fosdick and Phillips Mountains. I recognized that our team needed to cross that glacier, over ten miles wide, on snowmobiles and sledges. I looked at the photos for places to camp and to locate routes to reach various outcrops. I searched for a location where a Hercules could land and leave us, then come back to get us weeks later.

I saw evidence in the photos of strong, persistent winds. The prevalent gales had carved the snow in long rows of ridges trending northeast, corrugating the surface. I guessed these snow ridges to be a foot or more high and several feet apart—*sastrugi*, the Russians called these features. To land a plane here and travel on snowmobiles across that surface with sledges could be brutal, maybe dangerous.

The south side of the glacier in front of the Fosdick Mountains had thick windblown snowdrifts. In places, huge gullies or snow scoops guarded the sheer cliffs that faced the glacier. These looked a hundred feet deep. I reasoned that they formed from turbulent winds which beat against the range front. Large expanses of bare glacial glare ice, cleaned of snow by the wind, lay exposed at the foot of the Phillips Mountains on the north side of the Balchen Glacier, and on the south side of the Fosdick range.

"We'll need to travel across that ice," I whispered to myself. Crevasses were prominent in the glaciers that penetrated through the Fosdick Mountains and flowed down to the Balchen. I saw shadows of snow bridges across fields of hidden crevasses on the glacier. You die if you fall into a crevasse. My hands got a bit damp, and I felt thankful for the gloves.

McMurdo Station, November 28–December 5, 1989

I stood on the Hercules flight deck behind our navy pilot, Karen; her short blonde hair shone in the sunlight. We flew east toward our objective, the Ford Ranges, three hours away, where we would do a reconnaissance flyover, a Recce. Out the cockpit window over her left shoulder, the immense Ross Ice Shelf passed beneath. The size of Texas, made up of coalescing glaciers flowing off the continent, it floated in the cold, black sea.

Brightness overcame my eyes, blue and white in front and to each side. For a moment, I stared ahead, lost in the idea that I would be able to make out sights I could recognize. Before us lay a white ocean of ice with no boundaries or borders, with an emptiness that caused my head to hurt. The scale of nature here overtook me. I saw sights strange, immense, and unknown in my experience.

Chris had pulled us from our Survival School session a few hours earlier. She said that an all-women crew would make the Recce flight. Karen and her copilot, engineer, and navigator were young women navy officers. Cool. Karen looked attractive—tall, sandy blonde hair cut short, blue eyes, with an athletic posture and a friendly, easy smile. She was thirty-something, and me on the downslope of forty-something. Karen and I were both single. Around her, my memories of Annie back home dimmed.

"What do you do here?" I had asked her a few days earlier over wine in the McMurdo O club. We found ourselves on adjacent stools at the bar.

"I'm a heavy equipment operator. I fly Hercules," she joked with a big grin. That got a smile back from me. "I'm stationed at Point Mugu, near Santa Barbara," she said. We chatted about a bike ride together when we both returned. "Where are you going?" she asked.

"To do geology in the Ford Ranges, in Marie Byrd Land," I said.

"Don't know it, don't think we've ever flown out there. Going to camp out?" She smiled, seemed close to a chuckle. Did she think this sounded like a crazy plan, to go into Mighty Bad Land?

"Yep, six weeks."

"What's out there that's so interesting?"

"We're going to study the history of those mountains—how, when, and why they were created." I hoped I wouldn't lapse into too much detail and then be met with a glazed-over look or eyes that wandered and kill the conversation.

"That seems cool. Why is that important?" she said. Uh-oh. She asked a simple, tough question. I took a moment—*don't blow it.*

"It's part of a plate tectonic problem. New Zealand and a mammoth continental shelf around it were once joined to this part of Marie Byrd Land.

"You mean in Gondwana?"

"Uh-huh. They drifted apart sometime one hundred or maybe seventy million years ago. That created new seafloor of the Southern Ocean. Sounds simple, but the New Zealand side sunk below the sea to make a huge continental shelf and the Marie Byrd Land side went up and made mountains. Both sides should have done the same," I said. *Don't lapse into a lecture, professor.*

"Does this help to find oil and minerals?" she said. *Yikes. A tough question.*

"This is basic science, to figure out how plate tectonics works. Might help to find oil somewhere, but we're not looking. It's for knowledge, basic science."

She looked me in the eyes. "You're lucky you get to do that," she said, then smiled. My chest warmed, not from the alcohol. *Oh boy, now I'm hooked. She's interested in my work. Maybe other folks are too. How great.* But I didn't have time to follow up on this much promise—too bad.

Now I stood near her on the flight deck. Everyone on deck wore headsets so we could talk but all could listen. I didn't want to let a chance slip by to be extra friendly but couldn't flirt on an open mike. I felt awkward in the scene and around a woman as exciting as her.

I pressed the talk button. "How come all the crew are women, even the loadmaster?" I wondered why all women; why not have some women in all crews? She looked at me. The young woman navigator and woman engineer turned their heads my way. They had overheard the question. I felt stupid for a moment.

"The navy won't let women fly in combat zones," she said, "like submarine patrols along the coast of the US." That the coastal US was a combat zone had never occurred to me, but I kept a straight face. "For women to get in some challenging flying, they send us to Antarctica where it's a bit safer. We won't get shot at." She looked at me and grinned to share this ridiculous fact. I couldn't imagine flying out of San Diego on submarine patrols would be anything like the proven dangerous flying to, from, and within Antarctic skies. Most of the deaths in Antarctica these days are from aircraft crashes. I kept my thoughts to myself on this nonsensical rule of the US Navy.

We arrived in the northern Ford Ranges to find cloud cover obscuring much of the ground. I could see gray peaks and ice in gaps between wispy, low clouds—my first personal view of the location where we would work. In the dark shadows and mist, the scene looked cold and forbidding.

We couldn't drop below the clouds because an uncharted or mislocated mountain might be in the way. So instead, we continued farther east another hour to the Executive Committee Range to do a Recce for other scientists on board. Afterwards, we needed to refuel at Byrd Surface Camp, an outpost about four hundred miles east from our planned work location. They would be our nearest neighbors.

By this time, a full overcast had developed. No mountains around Byrd, so Karen dropped us lower below the cloud deck. We approached what looked like tiny black squares in a gray monochrome canvas. The whole scene looked like milk. Where would we land? Then lights came on to outline a skiway—we descended, touched down, and slid on our skis to a stop. Karen turned the Herc around and headed back to the black boxes. She cut the engines to idle, and a crewwoman opened the entry door and yelled at us, "Karen says you can go out while we refuel." I did. Hostile cold wrapped around my body. I shivered a few times.

I saw that the black squares were huts three-quarters buried in snow. Tall antennas sprung from their roofs, and steam blew from pipe chimneys sideways in a nasty, stiff breeze. We had landed on the featureless West Antarctic plateau. Here, mountains and valleys of the continent were buried under thousands of feet of ice that hid all hints of relief. Flat gray snow stretched as far as I could see in every direction. The gray sky and snow merged in the distance and obscured the horizon. *Man, this place sucks. A gulag. Are people sent here as punishment?* But I was overcome by the unique aloneness of the spot where I found myself. Not everybody had the chance to see what spread before me.

A single human figure about fifty yards away came toward me, dragging a fat fuel hose that looked stiff, frozen, and heavy. This appeared to be a struggle for the small person who dragged it. I approached.

"Need help hauling that?"

"Stand clear," replied a tiny female voice from inside a brown parka and behind a facemask and goggles. She leaned forward, dug her feet into the snow, dragged the rigid, thick, crooked hose toward the plane. She hooked up the hose, returned to the pump station, and started to refuel our Herc. I realized the rules were different for women in Antarctica. They don't need or want help from men.

It turned out that a couple of days later, I saw Karen at a barbeque in McMurdo attached to a handsome guy about a head taller than me with a long black ponytail—clearly a lucky Beaker. Oh well. I also squeezed in a visit to sick bay for my persistent groin pain. "I think it's a pulled tendon, maybe tendonitis," the navy doctor said. "Rest and don't lift anything for a few weeks." That would prove impossible.

December 5, 1989

The second Recce flight to the Ford Ranges happened after repeated days of bad weather in MBL. We had a new aircrew.

"So, Bruce, you're the PI, right? What's the mission?" Fred asked.

Our team met with navy pilot Fred and his copilot, Joe, a recent UCSB grad. Joe told me it would be fun to fly with a professor. He looked like one of my students: young, eager, having a good time—a steady grin on his face. His navy buzz haircut would've stood out on campus. Fred had hosted Tucker, Chris, Steve, and me for a tour of a Hercules the previous spring at Point Mugu Naval Air Station near Santa Barbara. That's where VXE-6 was based. I could tell then he was a welcoming and confident guy, and he was into our project.

Our team and the pilots sat at a table in one of the charmless, windowless rooms in Mac Center. A room with a distinctive military look, chipped Formica-top table, bare gray metal folding chairs. I brought a packet of photos and a couple of maps more detailed than those the aircrews used. I did the talking.

"Recce in the northern Ford Ranges, Marie Byrd Land. We'll be working there for six weeks, covering about twenty-four hundred square miles by snowmobile and sledge. We need to recon safe travel routes and camping spots," I said.

I passed Fred some aerial photos taken in the nineteen sixties. He shuffled through them. Did he appreciate my preparation? I couldn't tell.

"Where do you want the Put-In?" Fred asked.

"Here's a satellite image of a snowfield between the Chester and Fosdick Mountains—the Chester snowfield is about ten miles across. What do you think?" I slid the large-format, black-and-white print over to him. He pulled it close and inspected the image. He said nothing.

"Look good, Fred?" I wanted him to tell me that all looked perfect, no problem.

Fred nodded over the print, his hair dark but thin, a man in his late forties, my age. I saw the gold insignia on his olive uniform; I recognized his rank, commander—that was senior. Good.

"Looks soft and fluffy from space," he said, amused. "We'll see if that's true when we land on Put-In a couple of days from now. Can't see the sastrugi from this altitude, but I bet they're there," Fred said. My stomach sunk. I knew they were there.

"We need to parachute our snowmobile fuel ahead of the Put-In, right on that snowfield," I said. "Can you do it?"

"Airdrop. That'll be fun. Hope you can find it later, after Put-In." I saw him smile at the print. I looked at Dave—no reaction. Not find the fuel? That'd be a disaster. Why was the project getting more complicated and riskier every day? I decided to stay positive. These guys knew what they were doing.

December 6, 1989

Fred flew us out to our mountains late on a clear, calm day. We took off from the McMurdo Sea Ice Runway and headed east. More than three hours later, after hundreds of miles of vacant snow and ice passed beneath us, the Ford Ranges came into sight. Fred dropped the Herc in altitude and decreased speed. He lined up with the Chester snowfield and started a low, slow pass for the airdrop. Four fifty-five-gallon drums of fuel sat strapped together on an aluminum pallet—gas for our

Bruce on tailgate of Herc during Recce flight.

snowmobiles. The pallet lay on rollers arranged along the length of the deck in the plane's hold. The crew had rigged the barrels with parachutes.

Three crewmen in harnesses approached the rear of the hold and hooked themselves to safety straps attached to the plane. One of them let down the tailgate ramp, parallel to the deck of the aircraft. This opened the rear—bright sky and mountains burst into view. *Holy shit. The plane is open to the sky.* The sound level inside changed from the hum of the engines to the roar of frozen air shooting past the open rear of the Herc at almost two hundred miles an hour. *Who's seen an airdrop before? Firsthand?* I wanted to shout. I glanced at Steve and Dave who sat next to me on the red web bench seat. They stared to the rear, riveted by the action.

The crew hooked the rip cord of the parachutes to the deck and pushed the bundle of fuel barrels toward the rear, almost to the edge of the tailgate ramp. They turned to look at a bank of signal lights inside the plane. I watched the lights—red, yellow, green, with the yellow one lit up. The green light came on. The crew slid the barrels out of the plane. In an instant, I saw three parachutes burst open and then disappear. Our team looked at each other with big grins, gave each other thumbs-up.

A crewman approached me. He yelled into my earplugs.

"Captain says you can go out on the tail ramp to take photos."

Wow! This is gonna be so crazy. I unbuckled, stood up. Steve looked at me, a question on his face. He hadn't heard the message and didn't know what I was about to do. I made my way toward the rear, my camera strap around my neck. I looked at the wide-open rear of the plane and the blue-white empty air outside—like a gigantic TV screen in a dark room. A couple of crewmen stood by the opening, silhouetted in the blue-white light. They turned toward me. I felt hesitation, my legs rubbery. My apprehension came and then went.

The crew slipped me into a harness. I clipped on the safety line. I fingered the line—a single web strap about an inch wide. *Really?* My knees vibrated. *Is this a good idea?* The frozen air touched, stung my face. I sat on the tailgate at the edge of the plane. The air raced by. I heard its thunder, nothing else.

The violence of the air, the peace of the blue sky and white snow felt dream-like. Camera in my lap, I stared out at the wild, empty scene. I thought I might inch close to the edge, dangle my feet, get a photo of my boots in midair. This scared the hell out of me—that I had this vision. If I did it, would the safety line hold or would the ferocious atmosphere rip me out of the plane? I'd tumble through the blue, rocket down to the white where I'd make a puny red splotch smeared over the ice. That's when my dog tags would be needed. I snapped out of my delusion.

With caution but exhilarated, I prepared to shoot photos of the ground to study in McMurdo. Over the ramp edge, I saw the shadow of our Herc slide over the snow, ice, and rock only a couple of hundred feet below us. I pointed my camera, snapped away, at pace with my rapid breath.

What I saw matched the Byrd photos I had seen at the National Archives. Fields of blue glare ice protected the base of the ranges—blocked the rocks from our access. How would we cross that ice, stay safe? I saw immense and deep snow pits—large enough to hold a couple of Hercs, if stacked like mating birds—on the upwind sides of cliffs, cliffs we needed to sample. Could we go down into these pits and get out? I saw fields of crevasses, not open but covered in snow bridges. They waited for us.

We'd travel over the snow through those fields, not fly over them. The crevasses would be hidden, hard to see. How could we find them, not fall in, not be killed? Was there a way to unwind the clock, go back in time, and not do this? So much was at stake: money, people who invested in us, and equipment, the safety of my team, even our lives, and results that mattered. I let go of that idea. I couldn't—I wouldn't—turn back. I'd need to KIT; I'd need to *keep it together*.

CHAPTER 11

Put-In

Don't be a show-off. Never be too proud to turn back.
There are old pilots and bold pilots, but no old, bold pilots.

—**E. HAMILTON LEE**, the First Civilian Pilot to fly the US Mail in 1918

McMurdo to Marie Byrd Land, Ford Ranges, Fosdick Mountains, December 7–8, 1989

"Here's the spot we've picked for Put-In," I said. I handed Fred, our navy pilot, the satellite image of the Chester snowfield.

"This is the same photo, same place where we did the airdrop a couple of days ago, right?" Fred asked.

"Yeah. We want the Put-In at the center—should be almost level ground, small chance of crevasses."

"I like that, but you can't see sastrugi, the snow ridges, from space. I noticed some on the Recce. They can wreck our nose ski. That's a bad day."

"Don't know about those, couldn't see many in the low-level air photos." I realized I'd just lied to him. Why? I saw sastrugi in the air photos and on the Recce.

Fred looked at me. He didn't respond. Maybe he knew I was bullshitting him.

"We need two ski drags five hundred yards apart to see if crevasses open up on either track. If not, we'll land in the lane between the drags." He said this matter-of-fact. I thought I saw a faint smile on his face. Did he have the idea this would be fun?

I'd heard about ski drags, high-speed skims along the snow to locate crevasses in our landing path. I wondered what would happen if we found crevasses.

"Sounds good, Bruce. See you guys tomorrow sometime; watch the schedule."

* * *

That evening I made my way to the O club for my last night of relaxation. I entered through the wall of red parkas on hooks, saw a mountaineer I had a meal with at the Galley a day or two before. A real OAE, he had a lot of Antarctic seasons under his belt. He had told me a couple of aircraft stories he either knew about or took part in. I saw he had company, but I took an empty stool next to him. I sat down, ordered a scotch. He turned toward me, to be friendly, or perhaps tired of his present companions.

"When're you goin' to the field?" He eyed my drink.

"Tomorrow. Herc Put-In to Marie Byrd Land—you know, Mighty Bad Land," I said. I tried to project humor. "Open field landing, not visited before."

"Yeah, that'll be exciting." He shot me a big grin, his teeth white against a clean shaven and sun darkened face that looked to be sixty something.

"Whad'ya mean?" I couldn't help myself.

"There are stories."

"Tell me some." I took a big gulp, gasped when the scotch burned its way down.

"Last year they finished diggin' out a Herc that crashed in Northern Victoria Land in seventy-one, lost a JATO bottle, you know those rockets strapped to the sides, during takeoff, broke the nose ski, D59 site—open-field takeoff."

"Almost twenty years ago? They flew it out?"

"Yeah. Started digging it out a few years back—buried in the snow. Big job. Another Herc crash-landed there in eighty-seven, part of the recovery project, and two guys were killed."

I took another big swallow. This drink would be gone real soon if the stories didn't improve. But he was on a roll.

"At Dome C, it was worse, you know, the same general area, same type, open field, but nobody killed, in the nineteen seventies," he said. "First one, Herc goes down, then a rescue Herc crashes, breaks a nose ski—now it's stranded, then a third rescue Herc comes in and crashes." He laughed and shook his head. It was funny when no one was killed, I figured. "But they managed to sort out that mess the next year. Pretty impressive; got all three planes back. Had to fix those fuckers out there first." He threw back the rest of his drink.

I nodded, raised my eyebrows. "That's gnarly." He ordered another scotch for himself.

"Hey, did you hear about the science party that aborted their Put-In?"

"No, what's that about?" I had to hear this nightmare too.

"Just happened—they landed in a big blow in the Transantarctics, at their Put-In spot, and one of the scientists gets to the Herc door, looks at the scene outside, and refuses to get out. He stands in the door paralyzed. He says, 'I could die here! I'm not getting out.' He didn't. Ha, ha, ha."

"Really?" I stared at the racks of bottles behind the bar. What else to say? We fucked up anything and the story would spread like lightning through McMurdo. I thought about what our team could find on that snowfield tomorrow.

December 8, 1989

Our team marched over the ice runway past spinning propellers. The Herc that would take us to MBL in a few minutes waited for us. A brilliant, clear evening, calm, almost warm, the kind of soothing weather that gets you hooked on Antarctica as if the other miserable days never happened.

Evening flight was okay, but I would've preferred midday or earlier. Flight schedules don't care about that. Sun never sets, so why not fly? Besides, the evening or night low sun angle gave better definition to the snow surface and shadows, providing better outlines to avoid sastrugi and crevasses.

I looked up to the cockpit windows and saw Fred adjust various knobs and switches. Tucker led the way up the short flight of stairs to the door, then inside the plane. Steve, Cain, Dave, and I followed. Chris was last. Dark inside—I took off my glacier glasses to let them swing on their leash. We filed over to the right side of the Herc and sat against the fuselage on red web bench seats, faced toward the center of the aircraft and a line of equipment bags. Our orange bags were stacked in front of us. I had one of mine between my legs. My eyes searched the pile for my second orange bag, the one with my asthma meds. *Where is it?* Reflexive panic attacked my gut. I saw a bag tag with my name on it. That was it—good.

The tailgate ramp was down, the hold still and cool—our breath fogged. Chris sat next to me. She grinned. I entered a surreal mood, stunned, thinking of the adventure that faced me.

"This is it, Bruce. Yay!" Chris said. Her joy thawed the air. I pushed through high anxiety. I measured my breathing, made certain that I let my diaphragm move in and out. Events better line up. Work needed to start when we landed and run smoothly. I knew we'd lose time to bad weather once in the mountains. No opportunity to work could be wasted.

I gave my camera to a crewman. He took a photo of the six of us— clean, beards trimmed on four men, sunburn on Cain's and on Chris' pretty face. All the team fresh, well-dressed, showing smiles—we looked prepared.

To the rear, I watched our loads inch their way up the ramp and into the hold. I looked beyond the cargo to the ice outside. A pile of supplies caught my eye—including two snowmobiles and five sledges.

The rest of our gear. Because we planned to land in a virgin snowfield that hadn't been visited before, the crew had loaded the Herc light and nimble, divided our load of gear into two flights. About nine thousand pounds promised on a second flight was scheduled a few hours after this one. Our flight had two snowmobiles and two handlebar sledges—a minimum to travel and do any work. Also, not with us was a drum of glycol, the antifreeze I needed to drill rock samples.

"Hey, that must be the rest of our stuff," I said to Dave, pointing outside.

"Yeah. We're looking good, ready to go."

"We're screwed if the rest of our gear doesn't get to us right away. That fifty-five-gallon drum over there is the glycol I need to drill cores. I need that to work."

Dave shrugged. "Man—yeah, we'll find out. But we have the fuel airdrop at least." I was glad Dave was confident. I hoped we'd find that fuel right away.

Crewmen in the rear gathered up ropes and web straps and stowed them. Just another day to these guys. A crewman started the hydraulics to lift the tailgate ramp. The smell of the fluid reminded me of that warm day at Point Mugu last spring. The gate whined shut. The cargo hold transformed to quiet shadows. The six of us buckled our seat belts. Everyone put in their bright yellow earplugs to prepare for the roar we knew was coming. Our ears stood out in the low light. Fred began to taxi the Herc. In just a few minutes, we were airborne. We gained altitude, then banked east for the Ford Ranges in MBL more than three hours away. Like Chris said, *this is it*.

From my seat, I looked straight ahead at six large yellow plastic duffel bags by my feet. Bags labeled in black uppercase with the word SURVIVAL.

A few hours later, we reached the Chester snowfield in the Ford Ranges. The fasten-seat-belts sign came on. A crewman came by to check us. He pulled hard on our seat belt straps to make sure the latches held firm and tight. Fred cut back the power. The Herc started to lose altitude.

Fred swept the Herc in a large circle over the snowfield between the Fosdick and Chester Mountains. He leveled off the plane; we dropped lower, approached the surface. I braced myself, waited for the ski drag I expected.

The whiteness of snow reflected on the underside of the wings. Light from the few portholes increased, brightened the cargo hold. A bump and a strong jolt said the plane contacted the snow. Deep, loud sounds of *bam, ker-bam, boom, bam, boom, boom, bam* followed—a ride inside a giant kettledrum. The Herc collided with wind-carved ridges on the snow surface, sastrugi that Fred had worried about. Between booms and bams, the plane shook, creaked, rattled, squeaked, groaned, twisted in all directions from repeated violent collisions. *What's happening? Are we crashing?* My eyes were open, but I couldn't focus. I saw double, couldn't point my head in any one direction. After about a minute of skimming the snow surface, smashing into and through sastrugi, Fred applied power—the engines moaned, the collisions stopped—then we were airborne again.

What the hell! Was that normal? Across the interior of the cargo hold, a strapped in young crewman looked scared shitless. Our team didn't speak; we stared into space. My chest filled with wet concrete. We climbed. Fred banked hard and made a circle to allow a look down at the path of the ski drag for any crevasses that the Herc had blasted open.

The view must've looked good because he lined up the plane to repeat a ski drag off to the side of the first one. We descended to have that fun all over again.

Fred dropped the plane onto the surface once more. We skimmed for about a mile—same repeated collisions—then took off to make another circle to view the result. No hazards seen by the pilots and crew; Fred took us in for a third approach—this time to land. Maybe. I hoped the crew could tell if the drags had damaged our landing skis, like the nose ski. They couldn't know until we tried to land. My face began to sweat.

Our Herc touched the snow in the lane between the ski drags, endured jarring bangs and shakes before Fred slowed it near the southern slopes of the Fosdick Mountains. The plane halted. Props feathered.

Relief flooded me. Chris reached over to me, put her arm around my shoulders, gave me a big wet kiss on my cheek.

"Thank you," she said, then smiled wide. I wondered…what for?

Fred reduced the engines to idle and lowered the tailgate ramp. I could see out the rear of the plane at the bright, corrugated snow surface and the jagged, low profile of the Chester Mountains in the distance. The sun hung low in the evening sky. The snowfield appeared quiet. The scene looked benign. I felt reassured about this bit of luck.

The crew began to push our tall pallets of supplies down the tailgate ramp. Fred revved the engines, the plane moved forward, and the pallets slid out onto the snow surface. We would get out soon. Now that fact thrilled me. The crew did the same with more of our load pallets. Unloading happened fast. They meant to be rid of us in short order.

Cargo unloaded, we marched down the tailgate ramp and onto the snow. We had arrived. My first steps felt the hardness of the surface. My feet didn't sink in. I ignored this surprise. Preoccupied with what we needed to do, I didn't have time to grab the scene around me. If the cold lived here, I didn't feel it yet. We found a sledge among our loads with the crucial items for the next steps. Steve took possession of the large wooden first-aid box. It held splints and meds including morphine. We had learned to inject that in a Red Cross advanced first-aid class that Steve, Chris, Tucker, and I took in Santa Barbara. At the time, the idea of administering morphine to a team member had filled my heart with lead.

The team and I made our way beyond the gale of the prop wash behind the Herc, away from the powerful noise of its engines.

"They won't leave us here until we do the safety setup," I said to remind the team, but they knew this. I shouted above the engine roar. "Let's do our jobs we practiced." *We are ready. We're going to be fine.*

Cain and Tucker set up a Scott tent in about five minutes. Tucker started a stove inside it. He stuck his head out.

"All good," he shouted, and gave a thumbs-up. I checked that off my long mental to-do list. Dave and Chris found the sledge with our radio, the stranded-wire antenna, and three eight-foot-long bamboo poles.

Unloading Hercules at Put-In Fosdick Mountains.

The copilot pointed out the direction to McMurdo. The antenna needed to be set up at right angles to that direction; otherwise, our broadcast would be weak or lost. Steve and I walked off to string the wire up on the poles.

I placed our radio on the snow, then plugged in the antenna, powered the radio, and dialed in the frequency eight-niner-niner-seven. The radio equaled our lifeline. *Don't forget any steps.* My job now—establish radio contact with McMurdo. The final requirement before we could be left here. The copilot stood nearby and watched me. I got nervous. The radio had to work. I had to set it up right. I struggled to focus.

I picked up the handset and depressed the transmit button. "Mac Center, Mac Center, this is Sierra Zero-Seven-Zero," I said into the microphone, giving our code name. No answer. *Is the radio malfunctioning? Are they on a coffee break? Don't they know we've been dropped out here in the wilderness?* I remembered the chaos I witnessed in the McMurdo radio room; I sensed a wave of abandonment, but I needed to KIT. I repeated my call-in.

Hercules departure from Fosdick Mountains, Ford Ranges.

Then, "Sierra Zero-Seven-Zero, Mac Center, go ahead," came the reply in a muffled, distant, unconcerned tone.

"Mac Center, Sierra Zero-Seven-Zero, position seventy-six, thirty south, one forty-five, forty west. Six souls in camp, all is well, over."

"Roger that," came back, in a disinterested tone. "Mac Center clear." For them, this was no big deal. But I felt lightheaded relief.

I turned to the copilot and gave a thumbs-up. He acknowledged, strode over the snow to the Herc, then walked up the ramp and inside. Our team stood and watched the tailgate lift; the men inside disappeared. The pilots revved the engines. We turned away as the engines roared and a snow cloud blew past us. The pilots turned the Herc to the left. It lumbered its way back toward where we had landed, rocked side to side as the plane went over the snow ridges—looked like the wings flapped. In a short while, the Herc vanished into a swale in the snow surface.

Soon, we heard full power applied in the distance and an echo off the mountains. We looked toward where the plane had disappeared. The aircraft came toward us over the horizon, skimmed above the snow surface, gained altitude. The Herc passed low in front of us about a half-mile away, the slopes of the Fosdick Mountains behind. Chris leapt into the air and gave our Herc a touchdown sign. I watched it disappear into the dark blue eastern sky. A brown exhaust trail, evidence of its existence, soon dissipated. Our world became quiet. Silence. Now we were alone. We were here; we were set. I hoped the second flight would come tomorrow. Then we could hit the trail, begin to explore the Chester Mountains to the south, the Fosdicks north in front of us. Next year, we'd cross the Balchen Glacier and explore the Phillips Mountains. *This all would work. We can do it; we will succeed.*

I took a moment; I felt strong. I let myself grin. *Something's going to happen. I don't know what. We're about to discover, to create new knowledge. How great is that?*

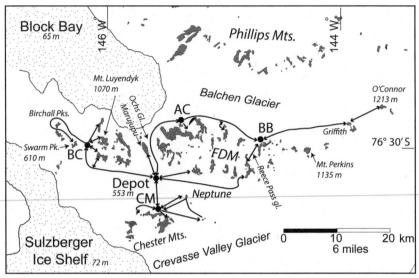

Map 5. *Close-up map showing trails blazed during the 1989-1990 expedition in the story. Depot was the main base camp. FDM is Fosdick Mountains. Camps made were CM (Chester Mountains), BC (Birchall Camp), AC (Avers Camp), and BB (Bird Bluff). Other significant locations mentioned in the book are Swarm Peak, Mount Luyendyk, Marujupu Peak, Ochs Glacier, Neptune Nunataks, Reece Pass glacier, Mount Perkins, Griffith, and O'Connor Nunataks. Several elevations are shown in meters (m). (After US Geological Survey maps)*

CHAPTER 12

Depot Camp

It was easy to see that here, nature was at her mightiest.

—**ROALD AMUNDSEN**, *The South Pole*

Fosdick Mountains, Depot Camp, December 9–11, 1989

I yelled. "Dave, Dave, wake up. It's snowin' in here."

Dave sat up in his sleeping bag. He pushed back his knitted, black beanie cap that covered his eyes and pulled out his earplugs.

"Huh? Hey!"

Both of us stared at the open tent entrance. Outside, a gray chaos reigned. Snow blew and drifted about, forced its way into our tent. A couple inches of snow covered our sleeping bags up to our thighs.

I struggled out of my bag; Dave too. I gathered up the tunnel entrance canvas and tied it closed.

"This snow will melt and soak our bags and freeze," I said.

Dave shook the rest of the snow off his bag.

"Forgot to close the tent tunnel before we turned in. Damn!" Dave said.

"Yeah, I thought I slept at the beach last night, so great outside—got fooled. A rookie mistake, big time."

"Whose idea was it to leave it open?" Dave said.

"Hey, we both liked the view—right?"

"We can't do that again."

"Okay, why don't you make sure we're tied up at night?"

"I guess I'll have to," Dave said, irritated as he continued to brush snow off his bag.

Neither of us grasped what could happen at any time: a change in the weather—for the worse.

Blizzards came without warning. Our team didn't get real weather forecasts because the nearest weather station lay four hundred miles away at Byrd Surface Camp. With no other data available, we radioed our weather observations to the Mac Weather office in McMurdo. I discovered similarities between our reports and Mac Weather's forecast. Dave had been up most of the night to call in weather observations every hour to aircraft control—the dreaded hourlies regimen. We hoped to get that second promised Put-In flight with two more snowmobiles and five sledges and more supplies. That flight cancelled. I began to doubt that USAP would follow through, deliver the rest of the gear we needed to start work.

Only once did Mac Weather warn us of a storm headed our way. Our team had been pinned down in a five-day gale. The storm faded. Everyone prepared to pack up for a camp move. Then I heard news. Almost ready to load our sledges, I reported our weather to Mac Weather and told them our travel plans.

Mac Weather thought they had a handle on a forecast. The voice on the radio sounded urgent, high-pitched, loud, and forceful.

"Sierra Zero-Seven-Zero, Sierra Zero-Seven-Zero, bad plan; hold position. Do not move. Hold position. Very high winds forecast for your area."

"How come they're warning us now?" I asked Steve. "First time they've done that."

"Barometers sinking again, winds now from the northeast," Steve said. "I think they're right."

"Let's go slow, pack up, but leave the tents up and wait," I said.

In about two hours, halfway through packing up, another gale clobbered us. We had to unpack our sledges, endure a blizzard for two more days. Mac Weather got it right.

My first inkling of what we'd have to face in an Antarctic blizzard was relayed to me by an OAE, Sean, who had come to visit me in Santa Barbara the previous spring. Older than me, Sean had blond-gray hair, was clean-shaven; he looked physically strong. Experienced, friendly, and generous—his help proved a strategic lucky break. We had to fill out a thick stack of forms, the Science Implementation Plan, or SIP, required by the USAP. The acronyms over time would multiply like a bad cold at the office. On the SIP, we listed everything we would need hundreds of miles away from McMurdo Station for six weeks—no resupply. Scott tents for living and cooking, snowmobiles and sledges for travel and moving camp, fuel for stoves and snowmobiles, cooking gear, food, sleeping bags, rock climbing gear, medical supplies, and so on.

Sean stayed at my home.

"Anything special you'd like while you visit?" I asked.

"I would appreciate a bottle of good Scotch whisky."

We started on that one spring evening—got through more than half the bottle. He'd been to MBL several times. We talked at my kitchen table.

"Be prepared for lots of tent days, Bruce," he said.

"What do you mean?"

"Storms, day after day; you'll be tent-bound. Bring books, music, games." He told me this cheerfully like these experiences were fond memories. Upon his news, I felt weighed down; this experience would be a grind. Tent-bound for multiple days in a storm? That didn't sound cozy.

"One storm, it lasted for a week, maybe more," he said. "One of my team developed a severe pain in his lower right abdomen."

No. Is he about to tell me an actual appendicitis story?

"I got on the radio with McMurdo medical—did a symptom checklist and it sure seemed to them that we had an appendicitis emergency. Not possible for a SAR flight; storm was brutal."

I waited to hear. Did this guy die?

"They asked if we had sharp knives, alcohol, and could anyone suture. I had a Swiss Army knife, real sharp, whisky, and a tent repair kit."

Holy shit.

"We were setting up for me to do it, then the weather cleared in a matter of an hour—called in a SAR, in a few hours got him out." He had a big smile on. "He didn't have appendicitis."

He wasn't making this up. Could we prepare for surgery? Struck me as impossible. *Just hope a bullet like that never comes our way.*

"Snow cleared up," I said. "Let's have some breakfast."

Dave opened the hinged top to the two-foot-square wooden food box that lay between our beds and dug around. He probed in the box, his hand clad in the dark blue polypropylene glove liners all of us wore. He rummaged through dehydrated potatoes, rice, beans, cans of stew, chicken, lobster, spaghetti, bags of trail mix (or Gorp), containers of Raro (New Zealand Tang), instant coffee, jars of peanut butter and jelly, hard salami, cheese bricks, candy bars—and found packets of oatmeal.

"Oatmeal for breakfast," Dave said.

He primed and lit the one-burner Optimus stove on the wooden shelf between us, careful not to melt his poly gloves. I sensed warmth. I grabbed a couple of snow blocks the size of my head from the space between the inner and outer tent walls by the tunnel entrance. We stored snow blocks for water inside the fabric wall on the left side of the tunnel and our boots on the right side. I put the blocks into an aluminum pot on the stove—piled high to resemble a snow cone. With the tent so cold early in the day, I put on a fleece jacket over my red turtleneck long john top.

Last night I'd gotten a bad chill, shivering without any control. I had put on all my clothes and gotten into my bag, drank hot tea, took about an hour to warm up. I hadn't put on enough layers. I thought about lying in bed with Annie, her platinum hair, the smooth brown skin of her back toward me as she slept. I touched her warmth, felt her breathe. But she wasn't here. I needed to take care of myself. Couldn't let a chill like that happen again.

The tiny stove warmed up the tent to freezing temp or even above. The snow blocks melted, sunk into the pot, converted into life fluid. The water heated to a boil. Dave dumped in the oatmeal, and I sniffed its aroma. Dave and I feasted—brown sugar oatmeal, my favorite—and hot Raro. I relaxed. Decades later, the smell of cooking oatmeal takes me back to this scene, repeated daily for weeks on several expeditions.

"Hey, Dave, you called in hourly weather overnight—right?"

"Yeah, miserable hourlies."

"Are we on the schedule for our second flight with the rest of our loads?"

"I asked. No. Had one flight slotted but got cancelled like the rest."

"We're gettin' fucked here, you know," I said. "NSF promised that flight two hours after Put-In. Now it's two days later with no promise."

"That really sucks, but man, yeah."

"This project can't be done with half the gear, and we can't wait here either, doing nothin'," I said.

"You've got an idea?"

"We can reach the Chester Mountains, make a move there, a short trip, and the Fosdicks are close. We can't overlook the Chesters—they're part of the story and part of our plan. Let's get the folks together, come up with a plan that'll work with just half our gear."

"You think Chris and Steve will want to go to the Chesters?"

"No."

I lay propped up in my bag with a cup of hot Raro. The wind buffeted the tent, but the snow stayed outside. I plugged in my Sony Walkman to the two bright yellow battery-powered speakers, the size of clasped hands, that I'd hung from the clothesline in the tent peak. We enjoyed progressive jazz at high volume, loud enough to hear above tent flapping and snapping. Dave liked the music but said, "How about playin' one of those weather reports you taped. That'll be crazy."

"Yeah, great idea." I put in the tape, pushed the Play button.

"*Patchy fog along the California Central Coast will give way to a mostly sunny afternoon. Daytime highs will be slightly warmer than yesterday. Most areas will top out in the seventies. A few inland valleys in Ventura County will soar into the low eighties this afternoon. Beautiful weather expected at Santa Barbara beaches all week."*

"Hey, I like the forecast," Dave laughed. Me as well.

Breakfast done; I felt the call of nature. *Oh no, now?* A hole we'd melted at the tent entrance tunnel, between the inner and outer walls, was meant only for piss. Fill up your personal poly bottle with the skull and crossbones on it and empty it down the hole. Number two—do it outside.

"Damn, Dave, I need to go out for a BM. It's a blizzard."

"Ha, ha, ha. Bruce, send me a postcard! Ha, ha, ha."

This would be nasty. None of us had time or energy to build a latrine out of snow blocks, and we didn't have a toilet tent. Could I wait until the blizzard stopped? That could be days. USAP had warned us about constipation, a problem with field parties waiting out blizzards. That would make my situation worse.

"Can't wait," I said.

I steeled myself, put on three layers of clothes and my boots, stuffed TP in my parka pocket. I untied the canvas tunnel and started to crawl out through it. Dave lay propped on one elbow, sipped his Raro, looked at me, shook his head, chuckled.

I pushed open the tunnel, crawled into it, and stuck my head out into the blizzard. I saw ferocious white and the dim shadow of Chris and Steve's tent. Commit. I jumped into the snowstorm. Wind blew horizontal and pushed me sideways. From inside, Dave grabbed the loose tunnel fabric and tied it closed.

"Good luck, Bruce. If you're not back in a couple of days, we'll search for you," he said. I heard him laugh loud, loud enough to hear over the wind.

Could I find a windbreak around, close? I knew that answer—no. *I'll try the downwind side of the tent.* I stumbled around, found a shovel, went behind the tent but found no peace. Snow swirled about with force. The wind and snow ignored the insignificant tent in their path.

Now I'd try my fancy drop-seat mountaineering pants. Suspenders attached to the front and each side, with zippers behind the suspenders, this marvel allowed a seat flap to drop without the need to pull down your outer pants. I made my commitment, dug a snow hole, took off my parka and wind jacket, then hung them on the shovel; they twirled in the wind. I turned to face the white blitz. Snow blasted me. I straddled the hole now filling with snow, unzipped my pant seat, pulled down my fleece pants and long johns, dropped down, and did the job.

A spray of ice crystals attacked me, like a swarm of angry Antarctic bees, stung my bare butt again and again. The wind tried to push me onto my back. Blasts of snow invaded my clothing. At first, I felt bitter cold; then I hurt. Would I get frostbite on my privates, become impotent? As quick as I could, I put myself back together, got on my jacket and parka crusted with snow, buried my evidence, and left the shovel to mark the spot.

Back at the tent entrance, I shouted, "Dave, untie, let me in. I'm freezin' out here."

I crawled through the tunnel, aware that my long johns and all else were full of snow. Dave greeted me with more laughs.

"Okay, Professor Numb Nuts is back." I slapped the snow off my parka and pants, shook off my balaclava, wiped my goggles.

Dave's response: "Ha, ha, ha—ha, ha, ha!"

"We'll see how long you can hold off."

"Remember to dig up your special present after the storm and put it in a night soil bucket. That needs to go back with us."

"I got the memo, Dave."

In a flash, I stripped naked, toweled myself, shivered and shook. I hung up my wet underwear on the clotheslines that ran around the tent peak tracing a square. Dave melted more snow on the stove; the tent warmed inside. I grabbed dry clothes and hurried to get them on as fast as possible.

"Dave, we gotta get it together. No latrine, leaving the tent open in a blizzard, me almost getting hypothermia last night, stuck here with only half our gear—things need to get under control."

"We'll be fine, Bruce. Be patient."

"Sure, but soon as this blizzard is over, first job is to build a decent place to take a dump."

CHAPTER 13

South to the Chester Mountains

Few men during their lifetime come anywhere near
exhausting the resources dwelling within them.
There are deep wells of strength that are never used.

—**RICHARD BYRD**, *Alone*

Depot Camp, December 12–13, 1989

"We need to find the fuel drop, ASAP." I stood with Steve, Chris, Cain, Tucker, and Dave on the snow outside our tents after dinner. "Could be anywhere on this snowfield, anywhere in a hundred square miles." I tried not to think about not finding gas for our snowmobiles or running out of the small amount of fuel we had while we searched.

"I should go with Tucker or Cain, start to look tomorrow, create a grid search," I said, uncertain of my plan.

Steve spoke up. "At Put-In, the copilot told me he saw the parachutes when they did the ski drags. They're northwest from here, maybe a few miles."

I looked in that direction, saw only vast white snow and blue sky. What if that advice was wrong?

"Hope that's right. Otherwise, we have a hundred square miles to search." I repeated the scale of the problem.

"Chutes are brown; they fill with wind. I can find them," Steve said.

Cain joined in. "I'll drive my snowmobile, tow Steve and the sledge with the barrel cradle. We'll go tomorrow morning, pick up one barrel at a time."

"We're counting on you guys," I said. I didn't remind them they couldn't fail.

"Piece of cake," Cain said.

Steve and Cain left to prepare the snowmobile and barrel sledge. Dave moved next to me.

"I don't think there'll be a problem," he said. "I've worked with Steve. He doesn't miss much of anything."

"Hope you're right." *Glad Dave's confident. We're on the brink of a special nightmare.*

"We're lucky we have gas for the snowmobile that came with us," Chris said.

"How's that?" I said.

"That barrel of MOGAS we have almost didn't make our Put-In flight."

"What?"

"I saw it sitting on the ice after the crew finished loading our flight. They left it behind."

"What the hell!"

"Yeah, I told a crewman to load it. He said I don't give him orders."

"I bet you straightened him out."

"We got the barrel." She pointed to our cargo line. I saw her smile, a smile that said she could get stuff done.

My head hurt to think that we almost got dumped out here with no gas, no way to search for the fuel barrels.

Early next morning, I heard a snowmobile starter cord yanked several times—lots of noise and bumps, then the machine fired up. Two men raised their voices over the engine rumble. Dave and I both lay in our sleeping bags propped up on one elbow, facing the pot of water that sat on the wooden box between us, eating our oatmeal breakfast.

"Steve and Cain are off to find our fuel," I said. Knowing Steve, he had a search strategy planned. Knowing Cain, he had to find the air-drop—it was personal. "Weather cleared, good day to search. I can see shadows of the shovel handles against the tent." I heard the snowmobile drive away.

"You'd think the navy would've set up some fancy locating beacon to find the drop," I said.

"You'd think, but not for us."

"We need to set up a work plan."

"Yeah, we'll be ready to start tomorrow if they get the fuel and the weather holds. That storm cost us three days." Dave poked the air with his spoon.

"I'm sure Steve and Chris have their own plan. Mine is that we start by going south to the Chesters," I said.

"Like you and I talked about. Yeah, makes sense."

"I figure we have maybe five weeks left before Pull-Out. That's tight. After Steve and Cain find the fuel, we'll get the team together to discuss what's next. But before we do any work, I need to figure out the magnetic deviation here, to adjust our compasses," I said.

Dave looked at me, puzzled. "Didn't the aircrew give us that?"

"I don't trust them on it. Also, we need to level the compass needles in our Bruntons. The south magnetic pole is close, and the north end of the needle will point almost straight up; that's useless."

"Yeah, like in New Zealand. Huh, hadn't thought about it," Dave said.

"Remember? Inclination is much steeper here, close to vertical. Need to take the glass cover off our compasses and slide the copper wire on the needle to balance it." Dave's face showed the mixture of thoughts that occur when learning crucial details that almost got overlooked.

Breakfast over, I put on my wind clothes, crawled out the tent tunnel into a bright blue day—calm. I squinted in the bright sun, put on my glacier glasses, stood straight up, stretched my spine back and forth, right and then left. The sun warmed my face. *This feels so good.* I closed my eyes, faced the sun, let myself sense the warmth.

Thirty yards away, Chris and Tucker worked behind one of the tents. They shoveled snow to dig a pit about chest-high to store our frozen food. Sunlight would thaw the food if left on the snow. Once finished, they would cover the pit with snow blocks to make a food cave.

I looked out at the Chester snowfield that stretched away from me in all directions, covered in new snow from the storm. I faced north toward the gray-black Fosdick Mountains capped by slivers of snow and ice—they looked close to our camp. To the west, a white snowfield carved by the winds spread down toward a distant ice-covered black sea beyond my sight. In the opposite direction, east, the snowfield formed the horizon in vacant space. Behind me, south, low purple peaks of the Chester Mountains broke the horizon into a sawtooth skyline. We'd go there soon if I could convince everyone. We needed to explore all the ranges.

This place, for me, felt so great—but where was the beach where I could put up an umbrella, a lawn chair, and have a beer? I grinned. Instead, our team could start work soon—a hopeful thought.

I heard a snowmobile some ways off. Soon it came into view, with Cain driving and Steve on a sledge. He sat on a fuel barrel that lay on its side. I pumped my arms into the air, excited with relief. They pulled up to camp and stopped. Cain bounded off the machine like a conqueror. Dave, Chris, Tucker, and I gathered around them, pleased. We laughed, excited that we'd jumped the fuel hurdle. I relaxed, grateful to dodge a fuck-up.

"We found the drop a couple of miles from here, down toward Mount Iphigene," Steve said. "Drove straight to them." I laughed at his certainty.

"Steve spotted it," Cain said. Now I knew what Steve could do— find his way and ours too. They unloaded the barrel and went back for the others, and the parachutes and aluminum air force pallet the barrels sat on when they were launched into the air. In a few hours, all fuel barrels stood on the snow in camp, huddled together like brown bears sitting upright.

Steve and Cain find fuel drop. (Photo: Steve Richard)

The fuel headache over, I began the search for the local magnetic deviation from actual north. I set up a sun compass and my Brunton compass on a rock box, squatted on the snow. I pulled out copies I'd made of the key pages of the current *Nautical Almanac* that would give me the true bearing to the sun at our location at different times.

Over the next few minutes, I took the sun and magnetic bearings to three peaks in view. At the end of a full hour, I'd calculated that the magnetic deviation at our location equaled eighty-eight degrees east—remarkable, almost ninety degrees. This enormous value in MBL showed me we were at the end of the Earth. Without adjustment to allow for this large deviation, our compasses would read north but in fact point us east. Our data would prove worthless. I almost laughed. First, I doubted myself, but I redid my calculations and got the same answer.

Satisfied and proud of my discovery, I announced the result to the team, who had gathered around. "I figured out the magnetic deviation; it's eighty-eight east. We have to adjust our compasses." I stood holding my orange field notebook with the calculations; I faced Chris, Steve, and Dave.

"Deviation is eighty-eight east. Unbelievable," I repeated. "The South Magnetic Pole is somewhere off to the west. We can't use our compasses, can't do our work, unless we adjust them for this deviation. If we don't adjust for it, we'll take bad data and could even get ourselves lost."

I couldn't see Steve's eyes behind his dark glacier glasses with the black leather side shields. He looked down at the snow, pulled at the sparse black beard on his thin chin. I imagined his mind at work: Is Bruce right; should I check it? Then he looked up. I saw that he decided to trust me. I then felt it safe to introduce another complication.

"And we need to know how to use the military grid navigation system too, that the aircrews use. We should keep in mind the important grid directions for communication with them. These ranges we're working in trend about east-west, but their grid trend is northwest-southeast," I said.

"I'll let you worry about that, Bruce," Dave said.

"We'll use compass directions for our data," I said.

I felt good about our successes and wanted to keep up the momentum of the day. Our camp sat tight now, food and fuel in place—missing two snowmobiles and several sledges, but work could start.

"Let's meet after dinner in my tent. We'll set up our work plan. We're ready," I said.

We tied open our tent tunnel to allow easy entry and a pleasant view outside. Our tent was a tight fit for six, so Dave and I rolled up our parkas and sleeping bags to create support to sit and lean on. Four teammates crawled on hands and knees through our tent tunnel and inside for the meeting. All of us maneuvered to sit on my rolled sleeping bag and on Dave's. We faced each other across the wooden boxes holding food and pots. Our heads and upper backs rested on the tent walls that tilted inward. I sat across from Dave; with his large body pressed against the tent wall, he cramped the space for Chris and Steve next to him. I began.

"We're ready to start work, so let's take stock. NSF approved our project to investigate the entire northern Ford Ranges—Chesters,

Fosdicks, Phillips Mountains. We planned to do Chesters and Fosdicks this year, Fosdicks and Phillips next year," I reminded them. "Dave and I talked this morning, and we want to start work south in the Chester Mountains. We can make a trip of a few days—no longer, because we only have two snowmobiles and two sledges. Don't have the two other snowmobiles and five sledges that we're supposed to get with that second Put-In flight—that hasn't been scheduled yet." I said the obvious, annoyed.

Chris and Steve squirmed a bit, then she spoke up: "We can do the same short trip north to Marujupu and Ochs Glacier, work in the Fosdicks," she said. "It's closer; we could even walk there, a couple of miles."

Steve backed her up. "We don't think there's anything of priority to find in the Chesters," he said. "The story will be the Fosdick Metamorphic Rocks."

"That's what Steve and I can do best—study the metamorphic rocks," Chris said.

I paused to think. Chris did a master's project on Arizona metamorphic rocks; Steve had also worked on them. Chris had visited Texas Tech to look at their rock collection from the Fosdicks. She had ideas of what to look for. She was the best prepared of all of us. I had to get her on board.

Dave spoke up—*good, some backup.* "We need to take a close look, can't dismiss the Chesters offhand without a look. You guys may be right, but we gotta do what we said we would—explore all the ranges," he said. "These ranges are basically unexplored except for the Fosdicks."

I trusted Chris' input, and Steve's. I needed to be strategic, make a decision that was sound. I took a minute. I didn't want to deviate from the plan we set up a few years ago, didn't want to waste effort either.

"We have to cover all the ranges," Dave repeated.

"How about this: we do a recon visit to Chesters, Neptune, Mount Corey, only a few days, say five, stay longer only if we find key structures. How about that? Then we focus on the Fosdicks," I said. "That would give us four or five weeks to work them. You guys can set up the Fosdicks work plan. In fact, I need you to do that."

Cain and Tucker had been silent, witnessing this conflict between the professor and his student without obvious judgment.

"Seems the Chesters is a manageable amount of work with the two snowmobiles and two sledges we have now," Tucker said.

"Gives us time to get the second Put-In flight and do the Fosdicks with four snowmobiles, six sledges, and all the gear we need," Cain said.

Not only was I relieved, but I was also impressed. Tucker and Cain's input made perfect sense. At this point, we settled on a decision. Chris and Steve relaxed. We boiled some water and shared tea and hot Raro. The next day, we would pack up to make trail and move to the Chesters.

Chester Mountains, December 13, 1989

The team stood together next to our two snowmobiles and two sledges at the base of a snow slope that led up to the low purple peaks of the Chester Mountains. Yesterday we'd traveled from the Depot and made a new camp about a mile away from this slope, a trip of five miles that took almost two hours. The drive was bumpy at first—a sledge tipped over. We crossed over fields of sastrugi. Closer to the Chesters, the northeasterly storm winds had drifted the snow to cover the ridges and the drive was smoother. We planted bamboo poles with red flags every quarter mile to mark our trail back to Depot Camp, in case of a whiteout. The trip didn't take long, but I felt exhaustion creep up on me. We needed to travel ten, twenty times farther to do what we wanted. That would be hard. To break camp and set up a new one tired me out, aggravated my tendonitis. Scott tents are bulky and heavy—and we had full wooden food boxes too.

I looked up the slope to the peaks. An overcast sky held back any clear shadows. The peaks seemed awash in gray, appeared to me a significant hike up that snow slope before we reached rock. "We need to rope-up here, in threes," Cain said. "Take your ice axes and put on crampons. Likely to be a bergshrund at the top of this snow slope against the rock. Need to probe the snow to search if we don't see 'em." *Bergshrund*, seemed a familiar word.

"Crevasses?" I asked.

"Yep," said Tucker. "Usually find them where the snow meets the rock." We formed two groups of three, tied onto the ropes with about forty feet between each person. We'd practiced in McMurdo, but here we couldn't make mistakes. Search and Rescue wouldn't chopper in. I didn't feel nervous though. Over time so far, I had watched and seen that Cain and Tucker knew how to keep us safe. I trusted them.

We walked straight upslope at a steady rate. In a few minutes, Tucker at the lead of my group stopped and pointed down in the snow with his axe. "Right here," he said. I panted, made my way up to the edge of the crevasse partly covered by a snow bridge, dropped to my knees, looked down—my first crevasse in MBL. The crevasse stretched across our path about fifty yards, hidden under a snow bridge beyond that. About two or three feet wide, it curved down out of sight. Beautiful blue light filled the empty slot. Quiet, lonely. I stepped over it with ease.

The group paused here. We untied and removed our crampons, then drank from a thermos bottle we each had. We scrambled up a rocky slope to the ridgeline. I sucked for wind. I glanced around—no one else breathless; my asthma again. *Can I keep up? They're all young and strong, leap from rock to rock like deer.*

I stood on the ridge, the rock of Marie Byrd Land, and looked at my feet, braced on purple-gray Ford Granodiorite, one of the major rock formations we expected to find here. *I finally did it*, I said to myself. I stood on the mountains of Marie Byrd Land, quiet with this happy fact.

I turned on the ridge, looked farther south to Saunders Mountain, the Denfeld Mountains maybe ten miles away across the Crevasse Valley Glacier, and the Swanson Mountains another ten miles beyond them—more than twenty miles. With almost no humidity or dust in the air, distant shapes of the mountains stood out. I looked at black islands in the white ice ocean, obscured some by low clouds and fog that started to drift in. Bright stretches of snow were dappled with the shadows of clouds; the peaks stuck out of the mists. The sight thrilled me. How many had seen this view? Not more than a handful of humans, maybe no one.

The south side of the ridge where we stood opened into a horseshoe bowl several miles across. The bowl faced the Crevasse Valley Glacier below us. I realized the bowl had been cut by a glacier when the ice stood much higher than now, hundreds of feet higher. I looked at the landscape, imagined the distant mountains disappearing, buried under a blanket of ancient ice that extended to the horizon. That vision stunned me.

The group broke up to explore, to work. We stayed within sight of each other. No walkie-talkies. We could signal with our police whistles, strung on a cord around our necks along with our dog tags and hand lenses.

I walked along the ridgeline over blue-gray granodiorite, came up to a black basalt dike sticking out of the rock a few feet and trending like an ancient, abandoned pasture wall across the landscape. I expected that these dikes might be here. Chris and Steve inspected some other dikes nearby, took orientation measurements with their compasses.

Down from the ridge, a white shape in a corner of rock caught my eye—a snow petrel, pure white, jet-black eyes, large pigeon size. She sat silently on a nest of pebbles in a wide crack a few feet away, looked at me. Above, a couple of others flew by. The bird on the nest vomited, a defensive move. I backed away. Open ocean was west of me, about fifty miles I guessed, maybe a bit closer, but a long flight for food. I realized there was life here—that made me happy.

CHAPTER 14

Neptune Nunataks

It seemed to me then that it would be almost impossible,
in this landscape, not to reflect on forces beyond the human plane.
Here, palpably, was something better than the realm of abandoned
dreams and narrowing choices that loomed outside the
rain-spattered windows of home.

—SARA WHEELER, *Terra Incognita*

Chester Camp, December 14, 1989

"Taking down the antenna? What's up with the radio?" Tucker asked.

He watched Dave and me bring down the stranded-wire antenna. We rolled up forty feet of silver wire braid around two bamboo poles.

"I'm taking the radio with us—gonna call Erick today, gonna ask him about our loads, why they're a week late," I said.

"Really, call the boss man—that'll help?" Tucker stared at me.

"We'll see."

"Be cool now, everyone will hear you." I sensed amused skepticism behind glacier glasses that hid Tucker's eyes, his sideways grin visible under his beard.

Sledges packed with two riders each and towed by our two snow-mobiles, our team got underway, heading east to Neptune Nunataks,

Chester Mountains from Neptune Nunataks. (Photo: Steve Tucker)

isolated low hills a few miles away. Soft snow, clear sky, and a sharp horizon plus good surface contrast made favorable conditions to detect snow bridges over hidden crevasses. The day felt comforting and calm. The snow was down-like, not bone-hard sastrugi that would rattle my vertebrae like old, dried peas in an empty soup can.

Three miles brought us to the edge of Neptune Nunataks—low tan-gray hills rounded off by prehistoric ice sheets of unknown age that had once ground their way across them to shape the rock surface, maybe more than once. Neptune, the name given to the blue eighth planet, cold, icy, and distant—an appropriate name for any place in Antarctica, but the wrong attribution here where rock the color of russet potatoes formed the surface. Neptune was the name of a geologist with a MBL survey in 1967, not the planet.

Our team pulled up to the outcrops, dismounted, and took off our red parkas. Warm enough to work in windbreakers for a welcome change. Machines shut down, silence prevailed, no breeze—so far so good. Ponds and puddles of solid ice spotted everywhere on the rock surface. Snow on the rocks melted in the sunlight; water flowed to low points where it froze into ice layers. I stepped on one patch; my plastic boots slid in different directions.

"Put on your crampons," Tucker said. Right, otherwise I'd do the splits, pull my other groin tendon too. I didn't need more of that.

"Thanks, man." I liked how he and Cain were looking out for me.

Boulders and rock cobbles lay strewn across the bedrock surface—glacial erratics, or rock garbage deposited by a moving ice sheet—all varieties scraped from hidden bedrock that lay deep under the snow surface and from unknown miles far into the continent interior.

The bedrock of the hills revealed a makeup of a complex mixture of rocks. They looked too chaotic to make sense. Reconnaissance maps marked Neptune rocks as Cretaceous Byrd Coast Granite, a common rock formation in this part of MBL. It's a buff-colored igneous rock with visible grains of clear quartz and yellow orthoclase feldspar, known to be about one hundred million years old—an ordinary rock crystallized from an ancient magma deep in the crust when T. rex was terrorizing North America. *This messy rock isn't Byrd Coast*, I thought as I scanned the ground. And it's too dark; it has different minerals.

Seemed that Mr. Neptune or anyone else hadn't been here. I concluded that someone assigned Byrd Coast to the rock in this nunatak by observations from a distance, maybe through binoculars, maybe from an overflight, not from a ground visit. Our team would become the first to step foot on this rare, isolated piece of ground. Too bad champagne wasn't on the lunch menu.

Steve, Chris, and Dave got busy, collected and labeled rocks, put them in canvas sample bags. Dave grabbed his sixteen-pound sledge-hammer and blasted the bedrock into large basketball-sized pieces. The whump and crack of the hammer were the only sounds.

Chris and Steve conferred over some samples they had knocked off the bedrock, looked at pieces through their hand lenses. I went over to them. "Hey, this bedrock isn't Byrd Coast Granite—even I can tell that," I said. "What is it?"

"Yeah, this rock looks like Ford Granodiorite, which is a couple hundred million older. It's not Byrd Coast Granite," Steve said. "It's maybe related to Fosdick Mountains rocks. We'll have to do lab work to find out."

"It looks in between Ford Granodiorite and the Fosdick metamorphics. Those maybe were made from Ford rocks," Chris said, "and this outcrop might be an example of that."

"Yeah, that would be a great story, figuring out where the Fosdick metamorphics came from. Maybe we've found a lucky spot that's a Rosetta Stone," I said.

I looked at my watch—time to make that radio call. Dave helped me string up the antenna; the team gathered around, attentive, anxious, and hopeful. I switched on the radio, then called Mac Center.

"Ready for the phone patch with the station leader, Mac Center," I said, and waited. I visualized the scene in McMurdo a thousand miles west. Navy kids in the radio room listening to loud rock, ignoring our radio comms, Erick with a cup of brewed coffee at his desk.

Erick came on the radio.

"How's progress out there, Bruce?" What? I couldn't believe his question. What did he think? We needed our gear, our loads.

"Slow," I replied, trying to stay professional and calm. "We don't have all our gear. The second flight that was supposed to follow us in a couple of hours never happened; been a week now."

"Can you work?" Erick ignored the blunder I pointed out.

"Limited. We need our second load: two more snowmobiles and five more sledges," I told him. "We can't travel far, can't cover enough ground to do our project."

"We've had to divert flights to the South Pole, Bruce, to keep up their fuel supply. Weather and aircraft issues crimped our schedule."

"Heard that before. We need to get on the flight plan, Erick." My pulse rate rose; my breath became rapid and shallow. I pulled my shoulders back.

"We're doing what we can; we need to supply Pole."

My forehead heated up. "Erick, our mission is compromised. You know this, right?" Silence. The team looked at me as though willing me to stay in control. All of Antarctica listened.

"We'll do what we can."

"We need our loads. Now, Erick," I said, on the edge, trying to keep my voice calm.

"We'll get you on the schedule, Bruce; be patient." I held back, took a few seconds.

"Sierra Zero-Seven-Zero clear," I said, and switched off the radio.

The others looked relieved that I hadn't lost control and told Erick to go fuck himself. I stared into the distance, breathed in and out, slowly. Alone, frustrated, forgotten, my good mood gone, all of us now abandoned in this absolute wilderness.

"Time is short, folks. This project is at risk," I said to the group. I glanced at Christine. She knew that meant her dissertation. "Time for a Plan B, guys. I'm open to ideas."

Ready to leave Neptune, Dave and Steve loaded the sledges with rocks we had collected—rocks that could reveal an important part of the story of the Fosdick metamorphics. Tucker and I took down the radio antenna. He pulled me aside.

"Hey, I need to switch tentmates."

"Oh yeah, what's up?" I said.

"I'm tired of sharing with Cain. He treats me like a rookie, acts like he's the big shot in charge." I noticed Tucker's mood had changed over the last week. Was this the reason?

"Oh yeah, he can be intense. Well, who do you want to share with?"

"Anyone will be fine."

"Okay. I'll talk to Dave about it." This problem wouldn't be easy to solve.

"Why ask Dave? You're in charge."

"Sorry, buddy. If you move, someone else moves too. I can't dictate that. Let me work on it. Trust me."

That evening in our tent after our dinner, I brought it up with Dave. We couldn't figure out a sensible rearrangement, and we needed the mountaineers to communicate as much as possible for safety's sake. They had to stay together.

* * *

We broke Chester camp the next morning for our move back to the Depot. Tucker, Dave, Cain, and I gathered around Dave's thousand-pound pile of rocks.

Dave addressed me. "Hey, Bruce. Steve, Christine, and I want to take a few hours to recon the west side of the Chesters on the way back to the Depot, a bit off our flagged trail. What do you think?"

"Off the flagged route? Should Cain and Tucker check that for slots?"

"Surface is dead flat over there—don't expect any crevasses," Cain said.

"Yeah, I agree," Tucker said.

"Okay. We'll need two trips to move the entire camp, but sure, why not; we don't have time constraints. Sun's not going down."

Clouds came in; the temperature rose on the thermometer attached to my windbreaker zipper. I saw a sky wall of gray snow in the distant east. *Maybe a storm's approaching?* I checked my watch barometer—looked steady but might be creeping down. Sledges loaded, our team drove toward the west side of the Chester Mountains. In about an hour, we stopped at the top of a slope where I could see the Crevasse Valley Glacier below. Looked like a steep slope, dropped maybe five hundred feet in a mile or two. First outcrop appeared in the distance about half-way down. Light snow fell.

"If we all hike down there and it snows harder, we could have trouble finding the sledges coming back," I said. Tucker and Cain looked at me—like this hadn't occurred to them, or they thought getting lost wasn't possible.

"We can follow our tracks back here; we'll be fine," Cain said.

"It could come down harder like last week—last a couple of days." Everyone paid attention to Cain and me.

"We're not going far. I think we can manage," Cain said. I wasn't convinced. Cain was bucking me. It seemed to me the others were with him.

"I'll stay here with the sledges; I already have my samples and I'm too beat from packing up camp to hike back up that slope," I said. "And I can signal with my police whistle to give you a bearing if we get a whiteout and you can't find your way." What I didn't say was my asthma was still a problem. Last night I'd puffed on my inhaler when Dave was asleep. Hiking up that slope would be hard for me.

"You can't stay here alone, Bruce. I'll stay too," Tucker said.

I watched the other four walk downslope. The snow fell harder; they disappeared into the speckled air. White flakes swirled down, up, sideways in a wind that was now hard. Snow obscured the gray profile of the mountainside. Flakes stuck to my face, melted, and made cold drips down my neck.

"Hey, Dave and I talked about your idea of changing tentmates," I said to Tucker, while I hunted for the hikers through binoculars.

"Oh yeah? What's the news?"

I faced him. "We need to keep it as is, need you and Cain to be talking safety, Dave and I planning, Steve and Chris designing their geologic strategy. Can you hang in there?"

"Really, four more weeks?"

"Can you try, for our project?" Tucker looked away.

"I'll try for you." He slapped my shoulder.

"That's great. Keep me posted." I knew he would, but still, I felt grateful.

Snow accumulated where Tucker and I stood. I tracked the hikers' progress with my binoculars. Tucker kept busy, checked the lashings on the sledges, adjusted the rocks into a secure position.

"This snow is burying their tracks," I said to Tucker. "Their footprints are shallow—surface is hard and windblown."

We both looked downslope, toward where the hikers left.

"I don't think they'll see us or be able to follow their tracks back here if this keeps up."

"Maybe. We don't know what the weather will be—could get worse, or better," Tucker said without concern.

"We need better."

Tucker and I waited. Cold now, I pulled up my parka hood, paced back and forth, swung my arms in circles to move blood to my hands and fingers, hopped up and down, looked downslope with my binoculars. The team and the mountainside had vanished in white. *Is a blizzard coming in?* How to tell? That would be bad, very bad; it'd separate the team. I tensed up.

I looked at my watch. Almost two hours had passed. I checked in the direction they left every few minutes. Considered what we would do if they couldn't find their way back—didn't come up with a good idea.

"What do you think?" I asked Tucker. We both looked south at swirling flakes and toward the now-invisible mountainside.

"I think it'll be tough for them to get back here," he said.

"This sucks. They don't have survival gear with them. Why didn't I ask them to flag their trail?"

"Too late now."

"Should I whistle for them?"

"Wait until they whistle us."

"Yeah. Then we know they want help. Think we'll have to go search for them?"

"No, that'd be crazy, and useless. We wait. Cain can handle this."

After twenty minutes, the snow let up some. Through a white-specked curtain, I saw two ghost figures pick their way back on bare rock less than a mile away downslope.

"I see two," I said, pointed into the gray.

"Two?"

I stared through my binoculars, straining for detail.

"Two more," I said.

"Nice. We're in the clear," Tucker said.

I took a few breaths to relax. About thirty minutes later, the hikers made it up the snow slope to the sledges under a light snowfall. By this time, all of us wore a covering of wet snow. We discussed what they'd seen. Not much new, they had a few samples, no surprises.

"Hey, we lost sight of you when that snow came in," I said to the hikers. "Thought you might have trouble making it back."

"How come? It was an easy hike in a snow flurry," Cain said.

"Could've been a storm, who knows? Next time we separate, I want you guys to take survival gear. I don't like taking a risk like this for so little payback."

"You worry too much, Bruce," Cain said. He marched away, slung his pack on his snowmobile, prepared to leave.

Not worry? Did he have a crystal ball? Nobody else showed any concern. The team could have gotten separated in a blizzard, with no supplies or shelter. Did I overreact? Was I too cautious? I didn't think so.

We mounted up and drove a short distance toward the trail flags, then headed for the Depot an hour away. Once there, we unloaded the gear and Dave's heavy rock pieces off the sledges.

* * *

The team made ready to put up tents at the Depot. The temperature now above freezing, snow fell and melted, made our clothes and equipment wet. My knees got soaked when I knelt in the snow to set up our tent. The shoulders and hood of my wind jacket got wet. My gloves soaked. Ice melted on my face. Cold misery crept inside me.

"We should go back to get the rest of the gear," I said to Cain.

"Okay. I'll lead in my machine and sledge. You follow with the other snowmobile," Cain said. Cain and I drove off; heavy snowfall appeared again. We followed the flagged route back to vacant Chester camp. The flags had been placed a quarter mile apart. That was too far, I realized; one-tenth would be safer. Passing one flag, it was difficult to see the next one in the white air.

At Chester camp, Cain and I loaded the rest of the equipment and supplies. "I'll lead back to the Depot," Cain said. He drove away. I followed thirty or forty yards behind.

Snowfall increased, harder, gray and white around me—a white-out. I lost sight of Cain, lost sight of the trail flags. I couldn't hear his machine. In a moment, I was alone. Had he ditched me? Now what? I stopped, looked down near my snowmobile and could see a faint shadow of the tracks of his sledge, about an inch deep in the soft, fresh snow. That guided me. I followed his sledge tracks, then passed a trail flag. Didn't know how he saw it; I couldn't until I was right next to it. Focusing on his tracks a few feet ahead, I followed them and arrived back at Depot Camp.

I shut down my snowmobile. I sat for a few seconds while wet flakes drifted around me, imagined myself inside a snow globe. What had I learned? Travel would be manageable, but I needed to adapt to unexpected challenges. I needed to respect my caution, not let go of it regardless of the opinion of the others. But we had to take chances to do our work. That was the job requirement. I had to trust that our team could do this, that I could do this.

CHAPTER 15

Marian, Ruth, Judy, and Punch

Travel when the sun is low, look for shadows in the surface
made by the sag of snow bridges over the slots.
My advice, avoid slots, go around, don't try to cross them.
If you travel across crevasses, you'll fall in and die.

—**PETE**, Antarctic Guide

Depot Camp, December 18–21, 1989

Milk. On my hands and knees, my head protruding from the tunnel of our tent, I looked outside at milk. I crawled out to stand, then looked about. All quiet in a whiteout, no breeze, air still. No shadows, no horizon, no snowfall. I read the thermometer on my windbreaker zipper—right at freezing, warmer than usual, not too much, good. No slush, no miserable wetness. Only a drab white.

I turned to look toward two orange tents to my right, then sledges and snowmobiles under tarp covers on my left. All suspended in milk. Weightlessness came to mind. I shuffled my feet, still stuck to the solid snow. I concluded that the tents and gear were on the ground also, not afloat. I put on my yellow lens goggles, no difference—no shadows. All man-made objects hung suspended in this Antarctic milk. I counted the sledges and snowmobiles. Off to my left, a box sat apart a fair distance from the rest of the gear, neglected. I didn't remember it. *Why is that there?* Did we forget to put a food box in our cargo line?

I decided to investigate, took a step toward it. *Twang!* A guy line of our tent stretched across my neck. The box revealed itself to be a piece of duct tape hung from the line. *So, this is a true whiteout. You can't trust your eyes.* I breathed a soft laugh. Then I crouched down to grab a couple of snow blocks to pass inside to Dave to melt for our water.

"You gotta see this, Dave, a complete whiteout," I said.

"Yeah, be out soon."

"Call in our weather—surface, horizon, nil, nil. As if they care," I said.

To my right about twenty yards, Steve exited his tent. He wore his dark blue long john top and bottoms, quilted booties on his feet. He placed two buff-colored Ensolite pads on the snow—the insulating sheets that lay under our sleeping bags. Housecleaning time? He bent down toward the open tent tunnel, reached for two silver metal pots that Chris handed him. The pots steamed. A bit of water sloshed out. Chris crawled out of the tent in her blue long johns, two towels in her hands. Steve and Chris stood on separate pads, then stripped naked. What the hell? I laughed; my breath steamed.

With washcloths, they dipped into one pot, then started to wet themselves, sharing a bar of soap and lathering their pale white skin, which contrasted with their deep brown weathered faces.

"Dave, get out here. You gotta see this," I turned inside our tent and said to him again.

"What's the rush? I've seen whiteouts."

"They're taking a bath out here, Steve and Chris."

"You're shittin' me."

"No, get out here."

He poked his head out, looked over—laughed out loud. Steve and Chris ignored us as they scrubbed. "I'm gettin' my camera," he said. Right, where's my camera? I fumbled in my pockets—where had I put it, with twelve pockets? Found the camera, took aim, snapped away. Dave joined me.

Steve and Chris took turns dribbling rinse water on each other from the second pot. Chris had soaped her hair. Steve rinsed it.

Chris and Steve moved fast—cold even if the temp held at just freezing. Dried off, they dressed and a minute later ducked back into their tent. This would be a good story. When did I last take a bath or wash up? I paused to count on my fingers back to our Put-In—fourteen days. Time to do something about that. I crawled back into our tent. Why clean up outside where it's colder?

Dave went outside to work on supplies. I dug into my gear bag to look for a package of wet wipes. They might be a good stopgap measure. Found them frozen stiff. On to Plan B. I melted snow in two large cook pots and got ready for a sponge bath. I defrosted a couple of wet wipes over the stove for use in my private territory. I ran my hands through my hair and beard—mangy, tangled, and a bit oily, but tolerable. I decided to skip my hair. I heated the water to a comfortable temperature.

I had brought three sets of underwear and long johns for the six weeks out here. About a week ago, I'd turned my first set inside out to wear them for another week; they must be done. I pulled out a new set from my duffel.

The temp rose; the tent warmed from the stove melting snow. I looked at my thermometer—almost fifty degrees. Stripped down naked, it didn't feel too bad. I stood up near the center of the tent, close to our single burner stove. Got a washcloth and soap. I worked fast. I could almost stand as I finished. Dried off, I searched in my gear for a deodorant stick—useless, too cold for anyone to have BO, but a nice idea. Found the stick, the top frozen on. I warmed the tube in my hands and my breath to get it open and wiped my armpits with the cold solid stick until it softened a little.

I put on clean underwear and long johns. Hey, that felt good. I might do this again soon.

I took the soap and rinse pots and poured the water down the pee hole at the tent entrance.

* * *

Before the whiteout, which lasted two days, we spent three days collecting rock samples and surveying the geology along the base of Marujupu Peak in the Fosdick Mountains a couple of miles north of camp.

"Marujupu—sounds like an American Indian name," Tucker said.

"It's made up. I read that Byrd named the mountain after three kids of Arthur Sulzberger, the *New York Times* publisher," I told him. "He funded a lot of Byrd's first expedition, the one in 1929 when they flew out and discovered this range. Marian, Ruth, Judy, and Punch, the son's nickname—Byrd took a few letters from each name until he had something that sounded sensible."

First visit to Marujupu, Dave and I rode on a sledge towed by Cain's snowmobile. Closer we got, the more I became in awe of that massif. The mountain appeared as a prow of a gigantic ship that sailed toward us through the ice, the sides of the mountain vertical, carved away by the Ochs Glacier that flowed downhill past it on either side. Broken low clouds rolled around the top, covered the peak every few minutes. Rock weathering had not happened to reduce those sheer sides. Elsewhere on Earth, we'd find rock debris at the base of the mountain that rested at the angle of repose. Here, solid rock rose ninety degrees straight up, no talus slope of thirty-five or forty degrees. The mountaintop showed itself flat and level, a plateau. Recon maps said the top elevation rose to over three thousand feet. The plateau topped a thirteen-hundred-foot cliff above the snow we traveled over.

Cain stopped his snowmobile before we got to the base of Marujupu. He dismounted and approached us, wanted to talk about a plan. His steps followed the track his snowmobile had just made. Dave and I watched him. He took a few steps, then he fell into the snow almost up to his chest—into a crevasse. Cain spread his arms across the snow to hold himself up. I stopped my breath—he hadn't roped-in.

"Oh shit," Dave said.

Cain looked at us. "Stay on the sledge," he ordered. He twisted his body left, threw his right leg up over the snow edge, then pulled up his left leg, rolled away from the hole, stood up, and looked down into it.

"What just happened?" I muttered. Cain walked back to his snow-mobile, got an ice axe, stepped across the open slot, then probed the snow with his axe and made his way to our sledge.

"That slot isn't wide; your sledge can get across it," he said. "Which side of Marujupu do you want to go to, west or east?"

"West," I said, guessing that he didn't want to talk about his near miss. I had to trust him.

He towed us across the snow bridge that covered the slot, then on to the base of the cliffs of Marujupu. Then I remembered the recon map showed crevasses on the west side of Marujupu but not the east—too late now. Why hadn't I checked that map before we left camp? Another mistake.

The next day, I teamed with Tucker. I planned to drill my first rock cores for experiments that would test the amount and timing of deformation in these ranges. Tucker inspected my rig.

"What's this motor?" he asked.

"It's a weed whacker modified with a chuck to hold an inch-diameter core barrel." I pulled out a few drill bits to show him—eight-inch-long stainless-steel tubes with tips of industrial diamonds.

"Hmm, this is a slick rig." As a contractor, Tucker liked tools and gadgets.

"I need you to pump water in that garden sprayer tank that's attached to the drill with this rubber tube—cools and lubes the bit while I drill." I showed him how the drill motor and water tank line hooked together.

"Amazed that we found standing pools of water yesterday," he said.

"Yeah, and deep enough to scoop up. A lucky break," I said. "Rock warms up from constant daylight and melts the ice that froze during the constant night of the winter months. I didn't expect that," I said. Just glad I didn't need to use glycol, that antifreeze, to cool the drill bit, which we didn't have anyway. A fifty-gallon barrel was on our second supply flight, the one that's two weeks late.

We walked up to the base of the cliff, carrying the drill and water tank. I looked up, leaned far back, craned my neck to see the crest of

Marujupu outlined against a mottled gray sky more than a thousand feet up. Looked to be more than vertical. It was an overhang.

Tucker and I stood in front of the rock face—waves and swirls of dark and light rock layers a foot across or thicker, a banded and contorted metamorphic rock called migmatite, all mixed up like marble cake batter—dessert for a giant with diamond teeth. This migmatite exposure was classic Fosdick Metamorphic Rocks. The lighter layers were granite and the darker ones made of biotite, a shiny black mica mineral. The rock appeared pristine, not weathered into clays, no soil. I marveled at it. Couldn't find rock exposures like this elsewhere on Earth—so clean and fresh, the rocks were not decomposed. Glaciers scoured away and left the exposure fresh—a geologist's dream outcrop, better than any roadcut. Not just at this spot but all along the cliffs that ran down the glacier for two miles or more. I loved Antarctica at that moment.

I stood up against the cliff, picked a granite layer to drill into, yanked the pull cord to start the motor, fired it up. The engine sputtered; I applied throttle. It whined, roared to life, puffed out a bit of blue smoke at first. Loud. The motor screamed and echoed through the glacial canyon. I pulled out ear protectors from my pack, put them on, tossed a pair to Tucker. This sound had not been heard here before—never such a disruption, intrusion, of the infinite peace we found here.

I raised my left foot against the cliff, put my left forearm holding the drill handle on my left knee. I aimed the drill with my right hand, increased to full throttle, leaned forward, and thrust the spinning drill bit into the cliff face—the rock screamed as if in pain. Tucker pumped water—spray flew in a fan around us. I held steady, arms and torso rigid. I leaned into the rock face, the drill and my body joined into one. I didn't waver, felt the rock yield to the drill that penetrated about six inches over three or four minutes.

It worked. Happy for that, I released the throttle to idle and pulled the drill bit out, taking care not to break the core still attached to the cliff. I looked at a perfect circle on the cliff face, a cylinder cut into solid rock. I shut off the drill.

Tucker inspected my work. "That's so cool," he said. "The rock cuts like cold butter."

"I'll orient the core now," I told him, and pulled out of my bag an inch-diameter aluminum orienting tube with a hinged platform attached to the top end. I inserted the tube into the cylindrical hole. Next, I placed my compass on the platform, leveled it, then read the tilt angle of the hole and its compass orientation. This was the step where the compass deviation correction I determined two weeks ago made all the difference. The wrong deviation equaled useless data equaled a failed expedition. I wrote down the core orientation numbers in my field notebook and gave the core an ID code.

I took a bare copper wire and slid it into a slot along the top of the orienting tube, scraped it in and out to leave a bright scratch on the core that still held fast to the cliff. I slowly pulled the orienting tube out of the hole. Then I inserted a copper wedge, like a screwdriver, into the circular gap I had drilled. I tapped the wedge with my rock hammer and snapped off the core at its base.

With my fingertips, I pulled the wet core out from the cliff and looked along its side. I saw the copper streak and let myself smile. I handed the core over to Tucker for inspection. "Look along its side for a bright metal streak; that's the orientation line I scratched with the copper wire," I said. He turned the core various ways, rotated it along its length, and smiled and nodded when he saw the thin metallic line and pointed to it.

"Orientation is important?" Tucker asked.

"Yeah, in the lab, we're gonna measure the direction of the magnetic compass frozen into these rocks."

"Frozen, like permanent? How's that?" he asked.

"These rocks were very hot when they were formed—look at the contortions in the light and dark layers, squishy like pudding."

"Yeah, I see."

I grabbed a marking pen and aluminum angle piece from my gear bag, laid the core along the aluminum edge, then drew a black line over

Bruce drilling rock core for magnetism research.

the copper streak and added feathers to one side to indicate the up direction.

"When they cooled, probably during uplift to the surface, magnetic minerals in them became magnetized in the direction of the Earth's magnetic field," I said. He showed his half smile that told me he got it, nodded his head.

Holding my drill, I paused and looked up again at the cliffs and sky to congratulate myself on the start of my work. I felt relief, reassured my part of the project would be a success. Now I had it, my first oriented rock core from Marie Byrd Land, the first of many others that I would take throughout these ranges during this expedition.

Near the top of the cliffs, I saw dozens of white snow petrels in flight, hundreds of feet above us. The infernal noise might have disturbed them—maybe not, but they had left their nests. They couldn't have liked the racket I made. Sorry I had to do that.

The birds circled downward, toward us. Got close, maybe a hundred feet above or less. Did they think we would harm them? Didn't appear so; they acted curious, innocent. They inspected us, the intruders—impossible that they'd seen humans before. They did not threaten us, but I imagined they wondered, *who are these creatures? Why are they here?*

The birds swooped around in lazy circles and figure eights, caught a bit of updraft against the cliffs, rose away and dropped down again—small white angels against the dark cliffs and gray sky. They had freedom here. Before us, no invaders had visited; no forces of other life threatened their peace. How perfect for them. We were free here too. Nobody to threaten us either. We were alone with them as equals. We shared this place no one saw now or likely ever would. Perfect.

CHAPTER 16

Pursuit of Happiness

Half the fascination an Antarctic expedition possesses is to be found in the sharpness of the contrasts experienced during its course, for it appears to be true that a hell one day is liable to make a heaven the next.

—**RAYMOND PRIESTLEY**, Geologist,
both Scott and Shackleton Expeditions

Depot Camp, December 21–23, 1989

Cain and I were in trouble. We landed on a pink rectangle—Entertainment. An ignoramus when it came to popular culture, I didn't know the names of top bands or the words to most pop songs. Over the years, as I listened to the lyrics of the Bee Gees hit song "More Than a Woman," I thought instead the words were "Bald Headed Woman"—I wondered, what the hell kind of song was that? Cain found himself in deeper trouble; as a Scot, these questions were North American—centric, and thus, a mystery.

Tucker held the Trivial Pursuit question card. Joni Mitchell sang low through my Sony speakers. He glanced at the question, showed the card to his teammate, Steve, who was crouched next to him on my sleeping bag. Tucker leaned back against the tent wall, smirked—it showed

First sight of Antarctica from Hercules flight deck.

Approach over Mount Minto and Northern Victoria Land.

McMurdo Station from Observation Hill, 1989.

The Billboard, seen on Recce flight.

Sarnoff Mountains in the Ford Ranges.

North front Fosdick Mountains, Ford Ranges.

The Team on Put-In day; Tucker, Steve, Cain, Dave, Bruce, Chris.

Unloading Hercules at site chosen for Depot Camp.

Mount Corey, east of Chester Mountains.

Birchall Camp and Mount Luyendyk.

Team geologists on Swarm Peak; Steve, Chris, Bruce, Dave.

Steve Tucker

Loaded sledges are prepared to descend the Ochs Glacier.

Mount Bitgood, north face Fosdick Mountains.

Steve Tucker

Dave ascending rope after crevasse fall.

Team travels up the Ochs Glacier to Depot in gale.

After Pull-Out from Depot; Bruce, Dave, Steve, Chris, Tucker, Cain

Mount Luyendyk, formerly Peak 1070, Fosdick Mountains.

through his black beard that thickened each day. Steve smiled and nodded. They knew we were screwed. They would win. Dave and Chris, the other team, looked on. The game had ended for them.

Tucker looked at me over his wire-rimmed glasses, asked us the Entertainment question. "How many lords-a-leaping are there in the carol 'Twelve Days of Christmas?'" Steve started the timer on his watch—one minute. Now Dave and Chris smiled too. Cain and I looked at each other, started to hum the tune in near synchronization. I mumbled "five golden rings," the easy part. Cain continued to hum, said "partridge in a pear tree" at one point. Yes, that's right, one of those. How many turtledoves, some other birds, milkmaids? I needed more time, pen, and paper. Cain said there were a few drummers involved too, and dancing. The other players started to chuckle. Yes, drummers, how many? Leaping lords, I had to guess, the timer almost out. "Eight," I said; the timer went off. "Wrong, need two more—ten," Tucker said.

Everyone laughed at the fumble we'd made. Steve and Tucker went on to win. I shrugged my shoulders in apology to Cain. He grinned and nodded. We needed to learn the lyrics—only five days until Christmas. I hung a sparkly green Merry Christmas sign outside our tent above the entrance. I felt the good cheer.

Later that day, Tucker had the idea to use the oppressive whiteout to our advantage. Without shadows, distances and scales were hard to judge. I had brought some plastic toy models of dinosaurs to hand out as team presents and prizes. Tucker took a triceratops model about four inches high and a companion palm tree, placed these down on the snow. He directed Steve to get into the background thirty yards away, near the tents, so the tree and the triceratops framed him. Tucker had Chris and Dave stand by their tents.

Tucker closed the aperture of his camera to get maximum depth of field, decreased his shutter speed—he had gone to the Brooks Institute of Photography and knew a few things. Steve posed so he looked like he faced the dino in terror, raised his arms in defense, leaned away from it. Chris and Dave took poses of flight and fight; Tucker lay out prone on the snow close to the triceratops. He snapped away.

When I saw the developed photos a few months later, it looked for certain that we had encountered the Lost World out in the white wilderness of Marie Byrd Land. A six-foot-high dinosaur attacked Steve at his tent door. Dave held a shovel ready to strike the dino from behind as Chris made her getaway. We survived Jurassic Park before Spielberg dreamed of that. I felt the warmth of that moment once again.

That evening, I met with Dave on the snow outside our tent. "I've got about two thousand pounds of rocks so far," Dave told me. "We should retro these on the second Put-In flight, or we'll have too much for one Pull-Out later on." We stood in front of our tents at the Depot. I looked at his rocks lined up, about two dozen, each piece the size of a large watermelon, headstones in the snow, a struggle to carry. Dave would grind these down to extract minute amounts of the rare minerals zircon and monazite. That's why he needed big samples. In his lab, he planned to measure radioactive decay elements in these minerals. That would tell their age—when these rocks formed.

Yes, I agreed, good plan. My paleomagnetism samples were small one-inch cores, no weight problem. Dave's would overload a plane at Pull-Out. We had weeks left for the expedition—several thousand pounds more could accumulate. I alerted Mac Center that we wanted to send back rocks collected so far when we got the second supply flight—that sounded good to them. Now we needed to get the flight, two weeks late so far, with maybe three weeks left out here. I prepared to turn in; my tension was building, but I hoped sleep would come anyway.

I pulled my arm out of my sleeping bag to check my watch: 6 a.m. Dave lay asleep. Time to call in weather soon. I slipped out of my bag, crawled over to the tent tunnel, untied it to look outside. Whiteout still. I looked at my thermometer, at freezing; read the altimeter, twenty-nine-forty. Pilots would use this to calculate their altitude above our camp. I ducked back in and decided to call the weather. Turned off the radio speaker so I wouldn't disturb Dave.

"Mac Center, Mac Center, Sierra Zero-Seven-Zero," I spoke into the handset. Dave stirred, rolled away to face the tent wall.

"Sierra Zero-Seven-Zero, Mac Center, go ahead," came the reply.

"Mac Center, Seven-Zero. Location at Put-In site, six souls in camp, all's well. Ready for weather?" I said.

"Go ahead, Seven-Zero." I read them the observations, finished with "Surface, horizon, nil, nil—Whisky-Five-X-ray." That should be obvious. Whiteout. No plane can land.

"Copy that, Seven-Zero. Standby for your aircraft events," they said. Huh? We were on the flight sched? "Seven-Zero, we have you listed for three events. Ready to copy?"

What? "Go ahead."

"Event A three fifty-four, launch twenty-two hundred, twenty-two December; alternate event, A three fifty-six, launch zero nine-hundred, twenty-third; alternate event, A three fifty-nine, launch fourteen-hundred, twenty-third."

"Copy that. You know we have Whisky-Five-X-ray here?"

"That's the schedule, Seven-Zero."

I signed off. Dave stirred; with his gloved hands, he pulled up the dark blue knit cap that covered his eyes. "What's that about?" he asked.

"Our supply flight is on for tomorrow. One flight, two alternates," I said.

He scoffed, then laughed quietly "Do we still have a whiteout?" he asked.

"Yes. I told them that. Whisky-Five-X-ray," I said. "Another waste of time. They believe they're trying, but they're not paying attention. First flight leaves at twenty-two hundred."

"That means we have to call in hourlies all night," Dave said. "We have to start at sixteen-hundred, four this afternoon. They'll get here at two or three a.m."

"Can you do the night weather; I'll do afternoon and evening? I've been havin' a hard time sleeping, don't think I could keep on schedule," I asked him.

"Yeah, okay."

After Dave and I had breakfast, I went to rouse the others and give them the news. They were outside to do various work on their gear. Most smiled when they heard it. "That ain't happenin'," Tucker said. Others shook their heads or swung them about to look at the complete whiteout that blanketed us.

Now still in whiteout conditions, immobilized, we spent the day writing duplicate copies of our field notebooks. We read, cooked, related, and discussed the geology and our plan. I wrote in my journal, "Feel bad about missing Christmas with my son Loren—miss him."

We all gathered in Dave's and my tent to have Mexican food—Tucker, the chef. The meal needed improvisation. "We're havin' burritos," he announced. He fixed Dinty Moore stew wrapped in tortillas, with some Tabasco sauce, rice, and partly frozen cheese sliced up with a Swiss Army knife—always sharp, those knives (I thought of Sean and his appendicitis story). Pretty soon, we had a Mexican food aroma in the tent.

"I feel like I'm downtown Santa Barbara," I told Tucker as I munched on mine. Ingenuity and a lot of time had put together a tasty brunch. Good idea to have a meal with each other once and a while—eat food more interesting than canned stew or canned boiled chicken with dehydrated potatoes.

I began the weather reports on schedule that afternoon. Dave and I had dinner, read our books, turned in.

Asleep soon, I stirred half-awake to see Dave crawl to the tent tunnel door. He untied the knot around the gather of tunnel fabric, stuck his head in the tunnel, flung it open with his right arm, stuck his head out. His large backside cloaked in dark blue long johns filled the entrance. I couldn't see what he was doing—only his dark blue butt. After he'd pulled himself back inside the tunnel, he got on the radio to report: "Whisky-Five-X-ray." I went back to sleep.

He called in the weather every hour. I slept through most of the reports. Near 1:30 a.m., I woke up while Dave slept. I lay awake and gazed at the tent walls. They looked bright yellow to me, like the sun had come out from hiding. I crawled to the tunnel, opened it, and stuck out my head.

Clear bright blue sky: the sun in the east faced our tent door. I stepped outside; my pulse pounded. I stood on the snow in my quilted booties and long johns, looking in every direction. Only glorious, open sky filled my vision. My chest expanded; cold air entered me, refreshed me. *Our flight will happen. That flight with our gear will get to us.* I ducked back in to tell Dave the news, then left to wake the others. The Herc should be here soon.

Over the past few days, we had written letters in anticipation that the second supply flight would bring mail in and take mail out. That would be a morale boost. My mother, Loren, Annie, and friends would be excited to get a surprise letter from the wilds of Antarctica. I needed letters from home. I hadn't been able to let go of the world. Hadn't yet been able to let Marie Byrd Land take hold and fully embrace me. I wanted my mail.

CHAPTER 17

Perfume on Ice

The English have loudly and openly told the world that skis and dogs are unusable in these regions and that fur clothes are rubbish. We shall see—we shall see.

—ROALD AMUNDSEN, *The South Pole*

Depot Camp, December 23, 1989

A Merry Christmas sign, green plastic with silver sparkles, a few inches high, hung over my tent entrance. The words flashed in the sunlight when a puff of breeze nudged them. We gathered outside our orange tents, stood on the snow in the bright sunlight, paced around, impatient. On the snow at the entrance to my tent, I placed our radio on top of a wooden rock box.

We saw a speck in the indigo sky to the south. A plume of brown trailed it—that identified the speck as a Hercules. I got on the radio and contacted the pilot, told him there would be tracks in the snow where the Put-In flight did ski drags, and landed about a mile away from our camp. Since then, snow might have covered some of the tracks. The team looked at each other, big smiles on our faces. We had sledges with two thousand pounds of rocks ready to send back. I held a

Bruce radios to Herc resupply at Depot. (Photo: Steve Tucker)

large envelope thick with mail ready to deliver. We anticipated getting a packet of mail for us.

"What about the surface, Seven-Zero?" the pilot radioed. He meant did the surface have sastrugi and how high, ruts how deep. I recalled the crash-landing experience of our Put-In flight.

I lied, "Some bumps, but not extreme." Why didn't I tell the truth?

The pilot banked the Herc, circled the snowfield. He dropped the plane into a glide to land. He didn't look like he was close to our Put-In landing spot, but rather farther away from us by a few miles. Why?

* * *

Mail, the link to the world, is a big event in Antarctica. The internet didn't exist there in 1989, and no personal phone calls left McMurdo. Handwritten messages on paper in envelopes with stamps made their way from Christchurch to Mac Town once or twice a week. The journey from America took at least two weeks—snail mail. "The flag's flyin' at MCC," someone in the Galley or the bars would report. We'd head up to that building, the Movement Control Center, for mail call, anxious, hopeful.

My cryptic postal address read: Bruce Luyendyk, McMurdo Station, Project S-070, PSC 469 Box 800, APO AP 96599–1035. APO stood for Air Post Office; I didn't know what PSC stood for. My mother complained about it. "What are these funny letters and numbers? There's only one Bruce Luyendyk in Antarctica. Can't they find you?" She complained more when I told her mail took at least two weeks for delivery and two weeks to return.

At the first mail call in McMurdo, I got a bunch of letters from stamp collectors. How did they get my address? I had heard of this—folks who wanted to get a letter mailed to them from Antarctica with its unique end-of-the-Earth postmark. Our project had a custom rubber stamp about two inches square, a cachet, made up with a design: FORCE—Marie Byrd Land. I stamped the letters, signed them, and dropped them in the slot at the MCC—felt good that someone cared to get contacted from Antarctica.

I had learned about this curious hobby at Woods Hole Oceanographic Institution in Massachusetts where I had been a marine geologist. I participated in an expedition with the submersible *Alvin* to the deep-sea hot vents in the east Pacific. After a dive to explore the sea floor at nine thousand feet down, we prepared to ascend.

The *Alvin* pilot flipped a switch to drop the weights that hung outside the submersible. Through my small porthole, I saw plumes of mud rise on impact, shining in the sub's floodlights. We began to float up. He opened a canvas bag and produced a handful of philatelic mail—from stamp collectors. "Here, we do this on the way up," he said. "It'll take us two hours to surface."

Three of us crammed into that tiny six-foot-diameter sphere, legs intertwined, leaned back against the cold titanium hull, and passed around a rubber stamp with the image of *Alvin* to mark the self-addressed letters. We wrote down our position and maximum depth and signed them as we drifted up away from the black, toward the pale blue, warm light.

I had given my Antarctic address to family, friends, my maybe new girlfriend, and my ex-wife. My former spouse, Jaye, sent me a photo. She held a sign that read "Antarctica or Bust" just below her ample bosom. "I'm stoked for you Bruce," she told me, which made me smile. A zany woman, but what did she mean? She had dumped me; it still confused me, still hurt.

My sort-of new girlfriend, Annie, appeared in my life right before I left; our relationship was new and had not reached an agreed-to status. She'd come to my house to look at a room I had for rent. Blonde, athletic, beautiful, with a magnetic personality; she looked like a dream. Annie sent me letters to Antarctica in scented pink envelopes. She had gone to Catholic school where she mastered perfect cursive penmanship. Annie revealed herself in spiritual messages filled with deep emotion. Unaccustomed to that intensity from a woman, I was confused by her letters. She followed astrology, a serious devotee. I scoffed at her for that—a grave mistake, soon retracted and soon forgiven. Joseph Campbell intrigued her, although I hadn't heard of him. *Follow Your Bliss* sounded too vague, the hero's journey and monomyth too evasive. I think differently now.

*　*　*

I watched the Herc's skis hit the snow and a white plume rise. It slowed down, moved along the surface, engines echoing off the mountains that bracketed the snowfield. The pilot spoke on the radio. "Negative on the retro…," he said with effort, voice wobbling as he tried to speak. We could see that the plane lurched and bumped over the rugged snow surface. He meant they would not pick up the rocks we had packed, and so would not pick up our mail.

"Say again," I radioed to check my own disbelief. The team looked on. No response from the pilot.

"What are they doing?" I asked the group. I looked at them—blank faces. Dave said the obvious.

"He means they won't pick up our rocks."

Steve said, "The ramp's down. They're pushing our loads out." Steve, with a hawk's vision, could see that far. I raised my binoculars and saw the tailgate down, our supplies shoved out as the Herc moved forward, rocked along on the snow. A colossal bird laying eggs while it waddled along.

I tried to radio them again. No answer—then again. I spoke on the eight-niner-niner-seven frequency. All of Mac Center and South Pole, the power centers, could hear this conversation and knew what just happened. Tucker reminded me of this: "Keep it together, Bruce. Stay cool."

We watched—tailgate went up. The Herc taxied, turned around, engines roaring, echoing off the mountains; it sped over the snow, tipped up its nose, and slid into the air.

"Those assholes," I shouted, held the radio handset by my side, did not press the talk key. I looked up to the sky at the exhaust plume left behind—it began to waft away and dissolve.

"Stay off the radio, Bruce," Tucker ordered under his breath.

We gathered around our sledges loaded with the rocks and a bag of mail for our friends and loved ones, dead letters for now.

Chris said, "Maybe the pilot didn't think they could take off with a load of rocks. Maybe he thought the surface was too rough."

"Oh yeah, how do they plan to pick us up in a few weeks?" I said. "We won't weigh less."

"Could be the pilot got pissed because you bullshitted him about the surface—not too bad, you told him. Not true; it's like landing in a field of boulders," Tucker said. Resignation set in.

Tucker asked for two to come with him and Cain to get the supplies, snowmobiles, and sledges that had just been dropped off. Chris and Steve joined them, speeding off to the west over the broad snowfield.

I watched them until they disappeared, but my thoughts were elsewhere. Tucker was right; the pilot was probably angry that I had lied to him about the sastrugi. Why didn't I tell the truth? Two times now—this pilot and Fred. I gazed at the endless white. I was fearful, I decided. Scared they wouldn't deliver our sledges and snowmobiles. We'd fail at our goals. I'd be held responsible for a waste of a great deal of money and effort. Unconsciously, I had traded the pilot's vigilance for my need to succeed.

In an hour or so, they returned, four snowmobiles and five sledges loaded with our long-awaited gear. I saw a white canvas bag with a zipper along its length—mail. I began to forget the fiasco that just happened. I imagined how good a letter would feel. I hoped for one or more for me, dreaded getting none.

I picked up the bag—one of us had unzipped it. "You're not gonna be happy, Bruce," Chris said. I dug my hand in, pulled out a thick manila envelope that felt full of letters. On its outside, written in black marking pen, I read, "S-161."

"This mail isn't for us," I said.

"That's mail for Vostok," Chris said, looking like she would laugh. I held in my hands mail for the Russian station in East Antarctica about two thousand miles from us. I opened the manila envelope. Maybe the packet got mislabeled.

I pulled out a fistful of aerograms, letters written on thin light blue paper that could be folded to make an envelope for international airmail. I read addressees with French names—return addresses for various locations in France, other European countries.

"What the hell?" I said.

"It's mail for the French scientists drilling the ice core at Vostok," Chris said. I lifted one of the letters near my face, closer to read the foreign names. I smelled perfume.

CHAPTER 18

Northwest to Birchall Peaks

Antarctica left a restless longing in my heart.... Its overwhelming beauty touches one so deeply that it is like a wound.

—**EDWIN MICKLEBURGH**, *Beyond the Frozen Sea*

Depot Camp, December 23, 1989

"So, good news, bad news, Dave," I said.

Dave and I stood on the snow in the bright sunlight outside our tent. We looked at the snowmobiles and sledges delivered to us by the Hercules a few hours before and the pile of rocks that didn't get loaded for return to McMurdo.

"Got our gear, but they wouldn't take your rocks. We'll have twice as much rock load at Pull-Out date." I shook my head, scoffed, kicked the snow with the tip of my boot, told Dave what he already knew.

"Man, took them over two weeks to get our gear to us. What happens when we get to our Pull-Out date?"

"I'm pretty nervous about it," I said. "We could be waiting a couple of weeks."

"I can't do that," Dave said. "I've got to get back to San Diego State to teach spring semester."

148

He had just started his job and didn't have tenure. He couldn't screw that up.

"Let's move the Pull-Out earlier a week," I said to Dave, "give us a time cushion."

"Okay. Yeah. Chris and Steve really won't like that," Dave said.

"Let's get everyone together in our tent and go over a plan," I said.

We met in our tent. I broke the news to the team. "Dave and I think we need to plan a longer window of time for our Pull-Out," I said. "We've just seen how tough it is to get a Herc out here, and we need more time to meet our schedule. One week earlier. To allow for delays," I said. "That will leave two weeks to work the Fosdicks this year. We'll make up time next year." Two weeks; that would be a disappointing outcome. Nothing could be done about it.

Dave nodded. "I have to get back to teach. I can't be a no-show."

"Also, we'll need two Pull-Out flights because they didn't retro our rocks today," I continued. "We'll be too heavy for one load. That makes it more complicated."

Chris and Steve looked at each other. She squirmed in her seat on my sleeping bag. I knew they would not be happy. Chris leaned forward to speak, elbows on her knees.

"Steve and I want more time; we need time," Chris said. "How about the second flight comes a week after the first?" She continued without waiting for an answer, "Steve, let's stay here for a second flight, get more time in the Fosdicks, maybe a week," she said to him. He looked up at the orange tent peak, stroked his black beard, grunted assent.

"Not without a mountaineer," I said. Cain offered to stay.

"Okay, I'll call in and ask," I said. Conflict settled for now.

Tucker changed the subject to our camp move today. We planned to travel northwest to Birchall Peaks to work out of a satellite camp. "Cain and I went over a travel scheme. We have all our gear now, four snowmobiles and six sledges, plus one spare." Finally. I let that relief seep into me. Now we would get to work with all our equipment. I wanted to get to the Fosdicks, to have Steve and Chris dig in and figure out the history.

"We need to travel to Birchall in two trains with two snowmobiles and three sledges each," Cain said. "Lead snowmobiles with Tucker and me, each followed by a sledge, snowmobile, then two other sledges—about fifty feet between snowmobiles and sledges. Last ones are handlebar sledges with Steve and Chris standing on the rear and braking," he said.

"Bruce takes middle snowmobile behind me," Cain finished.

"How do we rope-up?" I asked. We hadn't discussed this before. No need to—we didn't have the other two snowmobiles and four sledges and the spare until today. In my mind, I hadn't decided if we should tie ourselves to the snowmobiles and sledges or to the towropes. That was a decision for the mountaineers.

Cain answered. "We tie on to the towropes behind the snowmobiles, use Prusik knots. If a machine goes down a slot, we have a chance to jump off—don't get pulled in. Steve and Chris tie on to the towrope in front of their sledges." Sounded logical to me. Just hoped we never needed to test this scheme.

"What's your plan—camp locations and routes?" I asked Steve. "Remember I told you and Chris that I wanted both of you to plan that work." I had ceded these decisions to him and Chris. I figured I owed them payback for my forceful push to work in the Chester Mountains a couple of weeks ago, against their objections. I trusted them to come up with a smart work plan. The Fosdick range was metamorphic rock, migmatite. They were the specialists in examining those types of rocks. They could read the mineralogy and the textures that told the deformation history—the history of mountain building. Dave and I would sample in areas Steve and Chris identified as important to the story.

"We start in the west end of the Fosdicks, then work our way east up the Balchen Glacier," Steve said. "Chris and I think a camp on the snowfield between Mount Iphigene and Birchall Peaks would be a good central location. It gives us access to Thompson Ridge too. All outcrops are a few miles apart." He continued to describe their broad plan for a few weeks' work. "We figure five days' work there, then come back to the Depot, get more supplies, then head down the Ochs Glacier to the

Balchen Glacier and work the front of the Fosdick range to the east."
I nodded. Dave did too. The notion of travel on the Balchen with its
hidden wide crevasses gave me a chill, but I'd worry about it then.

I turned to face Cain and Tucker. "Do we need to recon the route to
the Birchalls?" I asked, concerned about crevasses along our path.

"No," said Cain. "Not much elevation change. Don't think we'll find
crevasses. Should be a bit downhill, fifteen miles, three or four hours.
Add a couple of hours to break camp and pack the sledges," he said.
Also, a couple of hours to set up at the other end, I figured—a long day.

"We'll flag the route every half-mile, to follow it back here in a few
days," Tucker said. That sounded too far apart, but I stayed silent.

"When do we want to pack up and move?" I asked to no one.

"Now," said Chris.

We had six sledges to prepare. "Each tent puts its food and gear boxes
on a separate sledge. One sledge gets a fuel drum for snowmobiles,"
Cain said to the group before we started to break camp. "One sledge
is for group food, but we divide some food and stove fuel between the
sledges." Right—if we lost a sledge down a crevasse, we wouldn't be
totally screwed. Chris had prepared our group food for ten days and
boxed it. Gratitude replaced my earlier burst of frustration at the pilot. I
felt lucky that Cain and Chris were competent, had planned, and knew
what to do.

"Tents stay up until we're ready to leave. They're lashed on top of the
loads," Cain said. "Doesn't look like any weather's comin', but that's what
we're doin' anyway." I liked that approach—not smart to assume. "Load
the radios on different snowmobiles—one with Bruce, one with Dave,
not on the first machines in the train." Of course not. A radio's not any
use down a slot.

I helped Cain pack one of the sledges. Nansen sledge designs were
more than one hundred years old, like the ones that took Amundsen to
the Pole. Made of ash, nine feet long, lightweight, and flexible with only
a frame and cross braces, they could carry up to a thousand pounds or
more. Teflon-clad runners were a modern update. Flexible frames made

for both good and bad. They could snake over bumps in the snow, the sastrugi, but they were prone to tip over if off balance—took two or three of us to right a loaded sledge.

We loaded wooden tent boxes built to fit on a sledge, strong but heavy, even when empty. Each tent had one food box and one box of pots. Chris and Steve brought over a share of the group food and fuel. On top of that, we placed our sleep kit duffels of sleeping bags, sheepskins, and insulated pads. Cain flung a long, box-shaped green canvas cover—called the tank—over the sledge load, now a bit over waist high.

By this time, the team had taken down our Scott tents and rolled them up, put them into orange canvas sleeves. This sledge Cain and I loaded belonged to Dave and me. He and Tucker put the tent and poles on top of the tank along with bamboo poles for the radio antenna. The tent looked like a giant orange beach umbrella ready for a sunny, warm outing.

To the side of the tent, we loaded a yellow plastic-coated canvas bag, big as a garbage can—Deep Field survival gear containing a lightweight mountain tent, signal mirror, flares and smoke bombs, dried food, and a stove with fuel. Each bag would support two people for three days—that was the theory.

Cain stood opposite to me on the other side of the sledge. We looked at the load. It appeared top-heavy to me, almost up to my chest, but he had done this before, many times, I reasoned, so I didn't question it. "Grab this rope and attach it to the oowks, pull tight, toss it back," he said, throwing a rope loop over the sledge load. *What?*

"What are oowks?" I asked.

"Oowks, oowks, down on the sledge," he repeated.

"Don't understand you," I said.

He came around, grabbed the rope on the tank and pulled it tight, slipped the rope under one of the stainless-steel *S* hooks attached to the sledge. "See, oowks," he said again, pointing to them spaced along the sledge, impatient. "You've done this before." He stomped away.

He meant hooks. I usually could make out what he said. I got steamed for a second. Yeah, I've lashed down a sledge load before but didn't get what he said through his Scottish brogue. I didn't like his attitude. *Don't make a big deal about it*, I decided.

I hooked the rope, pulled it tight, and threw it back over to Cain. He hooked it and threw the rope back again. We proceeded to crisscross the sledge and lash down the load. A couple of hours later, all sledges loaded were ready to go. We each tied our climbing ropes to our belt harnesses and the towropes and mounted up. Snowmobiles at idle, Cain and Tucker looked back at the rest of us. They wore motorcycle helmets, Cain's dark blue, Tucker's white. Driving in the lead, they were most likely to go down a crevasse, so head protection was a smart precaution.

All thumbs up, we got underway, at five miles per hour. My elation that now we had all our gear filled me with a rush of promise.

The trip over to Birchall Camp proved smooth. New, soft snow, sky clear and deep blue, made for a glorious trip. Undulations on the surface felt like ocean swells as I crossed over them. The new powder absorbed much of the roar of the snowmobiles. I had an easy time on my machine at the end of the rope behind Cain. His motorcycle helmet had a sunlit shine on it. Over my shoulder, I looked at the sledge I towed and behind it, the last one with Steve, who stood on its rear end, gripping the upright handlebars. His foot stepped down on the brake lever beam every so often to control speed when we headed downhill.

Our train skirted the west side of the massif that included Mount Iphigene. A few gauzy white clouds hung over its peak. They held my gaze. At that moment, I got in touch with my luck to be here. *I love this place and its secret beauty.* We followed the contour of the mountain's lower slopes and arrived at a broad snowfield a few hundred feet below the elevation of the Depot. Steve signaled to stop. This was the spot.

We made camp late, near midnight, unloaded sledges, set up our tents, made and ate dinner, turned in. I struggled into my sleeping bag. My groin hurt. My body vibrated. My pulse wouldn't slow. We had just finished a huge day. Gathering our gear from the second Put-In flight, packing up, traveling to a new camp. This was hard work. *Can I keep up?* The others were younger and stronger than me. No one else had asthma. I put in my earplugs, put on my eyeshades. I fell asleep.

CHAPTER 19

Swarm Peak

If Antarctica were music it would be Mozart. Art, and it would be
Michelangelo. Literature, and it would be Shakespeare...the only
place on Earth that is still as it should be. May we never tame it.

—**ANDREW DENTON**, Australian TV Producer

Birchall Camp, December 24, 1989

"What's this blue crap?" Steve asked. He squatted on a rock box placed on the snow at our new camp near Birchall Peaks. I faced him and looked over his shoulder toward the summit of peak 1070. The sun in the northeast to his back, he held a piece of light-colored Fosdick rock. Steve squinted through a hand lens magnifier, his face up against a chunk the size of his fist. A fine, clear day, no wind; we didn't need parkas.

I understood that he saw tiny blue mineral crystals in the rock sample and wondered what they were. Christine sat on another rock box next to his in the same pose. She squinted through her hand lens at another piece. They didn't wear gloves; it was warm enough not to need them.

The fingertips of their hands were wrapped in silver duct tape, like bandages, each digit wrapped in a piece of tape torn in thin strips. Their fingertips were cracked from dry air and dry rocks held in bare hands, no gloves. Steve rotated the specimen in his hand, touched it with steel gray fingers.

At first, we had worn thin, poly cloth glove liners, dark blue, when we worked with rock pieces. We hoped that we could handle the rocks in dry, cold air. Those gloves wore out. Next, we handled rocks with our insulated ski gloves, but the fingertips abraded away. Bare hands worked best to inspect rock samples.

I had a few cracks in my fingers too, but not as many as they did. My fingertips hurt like hell. Their fingers were in worse shape—I assumed theirs hurt much more. We had run out of Band-Aids and first-aid tape. When the fingertip cracks first appeared, we treated them with Zim's Crack Creme. That ran out and tips cracked anyway. Next, we used A+D ointment. Still, fingertips cracked. Duct tape held the cracks closed and protected fingers from added damage—less painful and more bearable.

"Yeah, I see those grains," Chris said. "Blue crystals."

"I think it's cordierite," Steve said.

"That would be so great." Chris continued to squint.

"Yeah, they've got garnet and sillimanite too," Steve said. "That would mean these rocks were metamorphosed at about twenty kilometers deep and at 700–800°C."

Twenty kilometers deep in the crust—twelve miles? And now at the surface? Stunning. I felt elated, happy for Steve and Chris for this significant finding. I looked at the mountains behind them. I imagined the rocks of the Fosdick range at rest deep in the Earth, ready to spring up. I thought of the scale of the event that brought the range to the surface. When did that happen? What caused it?

"But we don't know if it's cordierite until we get a look with a microscope," Chris said to Steve and me.

"Yeah, let's call it 'blueite' for now," said Steve.

"Blueite—love it." Chris scanned the sample with her hand lens.

I listened to them and realized what they spoke about. They were code breakers, could read the secrets of the rocks they held in their hands. Chris and Steve searched the samples of the Fosdick Metamorphic

Rocks for mineral clues to the history of the range. The minerals they found in these rocks showed they underwent extreme conditions deep in the Earth's crust. Now they were exposed in a mountain range.

"But when did the Fosdicks come up—what's the timing?" I asked them. Tucker wandered over and stood behind me.

"Need to get cooling ages," Steve said. "We need the time when these rocks got near the surface and cooled down."

"Yeah, cooling ages. You're gonna do the lab work?" I asked.

"Dave and I, we'll do that, get Paul Fitzgerald involved too. He could pin down the age of the last cooling." I knew Fitzgerald. He'd be a great addition.

"Why does that matter?" Tucker said. A consistently curious guy, he wanted to know why we went to all this trouble—to come to the Ford Ranges, the wilderness of MBL.

"It's about the process, how continents break up, how long it takes to get in motion, and determining what's the result. That's one reason we're here," I said.

"You're gonna figure that out?"

"Hope so." I looked at Tucker and laughed. Chris interrupted her sample inspection.

"We will, Bruce, for sure," she said.

"Okay. I guess it's settled."

* * *

Steve proposed a plan for today—Christmas Eve. Work would start late because we needed to rest up after our trip from the Depot yesterday evening. "It'll be good to do a circuit around the Birchall peaks." He pointed to them in the west. "Cross to the back side, see if the outcrops are good, look for structures," he said, "then come back to the side facing us." Fine, why not? Nobody had been here before that I knew of, so we couldn't miss anything new with a visit.

We organized our sledge loads for local travel cached food and fuel at camp, put survival bags on two sledges to take with us. We drove northwest a few miles to Birchall Peaks. The sun had dipped low in

Team prepares to ascend Swarm Peak. (Photo: Steve Tucker)

the southeast when we finally got underway—early evening, calm air reassured me. Blue shadows of the peaks stretched northwest, long on the snow. Our path circled around the north border of the peaks to the western slopes behind them. Snow on the peaks shone in the low sun and gave them a gold mantle.

We drove south, then turned east and climbed straight uphill to a pass between the peaks. The distance proved farther and the hill slope longer and steeper than I estimated. My snowmobile had it rough on the climb. Every day held a surprise. At the top of the pass, we stopped to talk about a work plan. The six of us stood on the snow, looked down a slope to Swarm Peak a few miles away to the southeast.

"All outcrops are snow-covered on the west side, not the best place to access rocks," I said.

"Yeah, I saw that," Steve said. "Better exposure on the east—storm winds blast the northeast faces clean."

"Swarm Peak looks bare, good to survey and take samples." I pointed to the southeast. Steve agreed.

We continued to the base of Swarm Peak and parked our four snow-mobiles and two sledges in its shadow. The sun lit up the far side of the range, outlined its profile with a silver edge. I took a photo of the dramatic scene. Near midnight now, we prepared to climb.

I looked up at the steep-sided conical peak. The recon map estimated its relief at six hundred feet. We worked to climb outcrops to the high points and then make a path down. We would collect samples along the way down so our packs ended up heaviest at the bottom.

The other five began to move up and I followed. The climb felt like a race. Everyone scrambled from boulder to boulder, no ropes needed. I arrived last at the top, winded—my asthma. I'd used my inhaler before I left, but I needed another shot. I shook it to see if its contents were frozen. Seemed okay. I sucked on it when no one was looking. I tried to disguise my breathlessness—couldn't appear weak. I didn't speak at first, kept my head down, pulled in cold air. I glanced over to Tucker. He looked to the distance, not at me.

Steve, Chris, and Dave searched the rock around the peak. They made observations, took notes, broke off samples with rock hammers, and looked at the split rock faces with their hand lenses. Cain bounded away to the east, hopped from rock to rock, catlike. Tucker and I stood on the peak.

My breath under control, I looked down at the bedrock by my feet—migmatite gneiss of the Fosdick range, the rocks that resemble a marble cake. I had suspected that from a distance as we circled the peaks. But the bedrock I stood on had a shine to it, a polish. I dropped to my knees and saw parallel grooves cut into the solid rock—glacial striations. The top of the peak had been polished and grooved by moving ice. How was that possible? I got out my Brunton compass. The grooves aligned northwest—southeast. *Wow, this is new.* The ice sheet had once overridden this peak, six hundred feet above the current ice level. The ice came from the southeast, not from the east as it does now. That was so cool. I laughed softly to myself.

"Steve, c'mere and look at these striations," I said to him, now close by me. Chris and Dave came up too.

"Hey, look at that!" Steve crouched down and started to take bearings of the grooves and make notes in his field notebook.

"The top of the ice had to have stood hundreds of feet above us to cut those grooves," I said. "Add that to the height of this peak, the ice must have been at least a thousand feet higher." I looked to the peaks in the distance and the ice sheet that padded up against them. The ice sheet had been much higher and thicker. Ice had covered those peaks too. Much more ice not too long ago had buried the landscape we now saw and lived in.

Tucker interrupted my thoughts. "Let's have a group photo of the geologists," he said. "Sit here on the peak. I'll get the Fosdicks and Balchen Glacier in the background."

We sat down, the four of us in a row: Steve, Chris, me, Dave. We faced Tucker and the sun. We pushed back our snow goggles and sunglasses. Chris removed her hat—her blonde hair showed a glint. I put my arms on my knees, pointed my ice axe down on the rock in front of me, toward Tucker. We all smiled. Tucker composed the photo. "Okay guys, everyone smile. This is fun—say 'ice.'" He took only one shot. The photo is on a wall in my home now. I see it every day. That image is a treasure from a special moment in my life.

I stood up; the others dispersed to work. I looked at the scene before me in the low light, the air so clear it couldn't be possible. No breeze, but calm and quiet. I gazed east and then south with the sun at my right hand. The beauty challenged my belief in what was real and I began to slip into a trance. I looked east once again. The gray cliffs of the Fosdick Mountains lined up to contain the Balchen Glacier; purple slopes of the Phillips Mountains constrained the other side of the glacier. Far away, the blue sky met a white horizon. How far? Had to be thirty miles. Beyond that point, an infinite land of ice—flat, empty, and white, without any life. The scale of the scene, of us lost in its presence, confused me. I couldn't make sense of this infinite emptiness.

South of me, other mountain ranges stood violet-black in the low light. I recognized some of them, twenty miles away for sure, maybe thirty. All empty—everything in my vision and beyond it. The beauty made me lightheaded. My body was invaded by the power of the moment.

I couldn't forget this time, this place, and this privilege. I hoped to see this vision, now before me, when I died. Could I make that so, wish for it, see this sight in my mind for the years to come so it was part of me? Not fade to black, but to the blue, the light, the white.

CHAPTER 20

You've Got Mail

*Sure, the lion is king of the jungle, but airdrop him
into Antarctica, and he's just a penguin's bitch.*

—**DENNIS MILLER**, Comedian

Birchall Camp, December 24–26, 1989

Christmas Eve, back from Swarm Peak, Dave and I had finished an after-midnight dinner of freeze-dried pasta. We heard sets of footsteps crunch the snow outside our tent. I guessed they belonged to Steve and Chris. If you're around folks every day for weeks, it's funny what traits you tag them with. The steps stopped at our tent.

"God rest ye merry gentlemen
Let nothing you dismay…
O tidings of comfort and joy."

"Christmas carols." I said the obvious. "How crazy is that?" I stuck my head out the tent tunnel to invite them in. "Hey. Thanks for the serenade. How about some Christmas cider?"

Our tent had some decorations. I had brought a paper Christmas tree, green paper with white tips on the branches—the kind that opened

161

like an accordion. We stood it up on the wooden tent box in the center of our space, enjoying our silly attempt at normalcy.

We shared cider with the smell of cinnamon and had some laughs as we talked about our unique circumstances.

"We thought we would climb peak 1070 tomorrow, on Christmas," Chris said.

"Climb it? That's about fifteen hundred feet up," I said.

"Cain will recon for a route, shouldn't be technical," Steve said.

"I'm not up for something like that," I said, realizing I would hold everybody back—short of wind means slow uphill.

"I'm in," Dave said. "Great idea."

"Well, I'll hang here and drill outcrops over at Mutel," I said. Work on Christmas. Why not?

Later, Tucker found out that I would not be going on the climb. "I'll stay here and help you, Bruce," he said. "Not a good idea for you to be alone. Field rules are that nobody can be separated."

On Christmas day, Tucker and I drove over to an outcrop east of us while the rest of the team started up peak 1070. Tucker appeared quieter than I expected, not with his usual cheer.

"What's up?" I asked.

"It's Cain, about the climb. He said everyone should hold back until he goes and checks out a route. He can't do it any better than me," he said.

"Still treats you like an amateur, huh?"

"Yeah, I'm not happy about it. I don't need him to check a route for me."

"Well, I trust you. Thanks for spending the day with me," I said. "I really appreciate it. I know you'd love to do that climb."

"Yeah, I would've."

I hoped Tucker and Cain could keep a lid on things until we finished—just a few more weeks.

Later we gathered in our tent to celebrate the day. Dave put on his Big Red to take the role of Santa Claus, but with a black beard. We sang parts of Christmas carols, didn't know all the words. Steve remembered a Boy Scout song to fill the void—"On Top of Spaghetti" ("Old Smokey").

"On top of spaghetti,
All covered with cheese,
I lost my poor meatball,
When somebody sneezed."

And so on. Spirits restored as with any good Christmas gathering. We turned in as a breeze began to flap our tents. Earlier, I had noticed a thick layer of clouds approaching. My watch barometer showed the pressure dropping. Light dimmed like an eyelid closing.

December 26, 1989

"I hear a Herc," Steve said. We all stopped and listened, interrupted our party chatter, six of us crammed into one tent to celebrate Christine's birthday, day after Christmas. I gave her a beefcake calendar, "Men of Summer." She felt moved, I think, that I carried such a trashy gift so far and gave it to her out here in this emptiness. We shifted our positions on the sleeping bags we sat upon—boot liners kicked others, knees bumped and touched, smiles showed all around.

"I don't hear anything," I said. UB40 performed "Red Red Wine" through the speakers attached to my Walkman. I turned it off. How could anyone hear a Herc inside the tent? Steve, though, could hear and see better than any of us, a kind of bionic geologist. I sharpened my ears, focused.

A few said they heard a Herc too; the distinctive moan of the Herc's unique engines. Why would a Herc be overhead at eleven-thirty at night out here in no-man's land? Since last night, a full overcast covered us—snow flurries too. This wasn't flying weather.

"Let's call Mac Center and ask what's up," I said, excited. I wedged myself between Chris and Steve, reached behind the food boxes, flipped on the radio, and pressed the transmit button to call in. I had charged the batteries with the solar panels before the weather started to turn.

Mac Center said yes, a Herc should pass overhead. It had stopped at Byrd Surface Camp to our east and now headed back to McMurdo. The crew had mail to airdrop for us. Mail delivery by parachute out here. Surprise, we weren't told of this plan. This could be so cool.

"Hey, they remembered us," I said. "Who would've thought?" A few laughed at my remark, but all showed alert interest, sat up straighter than a couple minutes ago.

The pilot came on the radio, a woman's voice. "Sierra Zero-Seven-Zero, we can't see the ground—full cloud, don't know where you are, so negative on the drop," she said.

Now concealed in clouds and snow, we'd missed the weather of opportunity—fine yesterday, of course. I acknowledged the pilot's message, then felt deep disappointment. I stared into the microphone to imagine them flying away. Clear and beautiful weather this morning. Christmas yesterday and Christmas Eve an inspiration, with clear, calm skies, and a view that stretched for tens of miles.

"Bummer," someone said.

"Yeah, won't be another shot at it for days, if then," I said. "But they're thinking of us." I bet to myself we wouldn't get any mail. *We are forgotten out here. We are not convenient.*

Balchen Glacier, another day, much later

One morning on radio check-in, Mac Center gave us a welcome message. "We have a mail drop scheduled for you tonight, maybe around zero one-thirty." I wouldn't get my hopes up for sure. Contact with the world would feel weird about now. But first we had a camp move to make, a trip of more than twenty miles over the Balchen Glacier.

Arriving, I struggled off my snowmobile near 1 a.m.—my butt sore, back stiff, hands numb—after twenty-four miles and four hours, relieved to be done. "Let's have burgers," Dave said after we dismounted and gathered around. "I'll defrost them in a pan." Nice, should be ready about 2 a.m.

"We need to set up the radio, give our position, check on that supposed mail drop," I said. We put up the antenna and our tents, gathered food, unloaded the sledges, stacked up rocks and sample bags.

I looked forward to a mail drop and frozen food. When departing sixteen days ago, we had overlooked cross-checking our food supplies—left frozen meats and vegetables behind. We'd eaten a lot of bulgur the last couple of weeks, a favorite for Chris, fine with her. Plenty of Cadbury chocolate and trail mix otherwise. We added variety with camp-made desserts that needed to be kept in a freezer—so easy when you live in one.

"Mac Center, Mac Center, Sierra Zero-Seven-Zero," I called into the radio.

"Go ahead, Seven-Zero," came back.

"Checking on the mail drop—still on?"

"Charlie that, Zero-Seven-Zero, X-ray Delta Zero-One headed your way from Byrd." A night of perfect weather persisted. They would fly over us if they could.

"We've moved camp, Mac Center," I said.

"Give us your coordinates."

"We're on Balchen Glacier a couple miles grid southwest of Mount Perkins, latitude seventy-six, twenty-nine, longitude one-forty-four, eleven—how copy?" I said. I'd given them our grid bearing, not compass bearing.

"Roger that. Lat seventy-six, twenty-nine, lon one-forty-four, eleven," Mac Center replied.

Steve listened; he interrupted me, "Better check they don't think you gave them decimal degrees." Yes, I recalled the chaos of the radio room—anything goes with those navy teenagers; if they got our position wrong, we wouldn't be found.

"Mac Center, to clarify, lat seventy-six degrees, twenty-nine minutes, lon one-forty-four degrees, eleven minutes; how copy?" I said.

"Zero-Seven-Zero, say again," replied Mac Center. They didn't get it. I looked at Steve; he smiled.

"Mac Center, I can give you decimal degrees—wait one," I said. "Mac Center, lat seventy-six decimal five-zero, lon one-forty-four decimal seventeen. How copy?"

"Roger, Zero-Seven-Zero, lat seventy-six, twenty-nine, lon one-forty-four, eleven," came back.

"Ha. They don't know what you're talkin' about, Bruce," Steve said. He was right—they had no idea where we were and didn't care.

"Zero-Seven-Zero clear," I said.

Steve said, "I hear a Herc."

"See it?"

"Nope."

The voice of a pilot came on our radio, relaxed, unconcerned. "Sierra Zero-Seven-Zero, we're over a glacier, we think, have your mail bag, but can't see you—what's your location?"

"I see them, maybe twenty miles down the glacier," Steve said, pointed. I saw them, Dave too; a speck moved in the sky, appeared to head away from us.

I called the Herc, "X-ray Delta Zero-One, we see you down the glacier, twenty miles from us; we are grid northwest of you. Looks like you're headed grid northeast."

"What glacier, Zero-Seven-Zero?" *They're lost. Our camp is too small, too insignificant to be seen. We need to signal them.*

I pulled out my Brunton compass with the mirror in the lid.

"I'm goin' to signal you with a mirror flash, Zero-One," I said.

"We'll look in your direction, grid northwest," the pilot said.

I stood, faced down the glacier toward the speck in the sky, opened my compass lid so the mirror faced the sun, low in the night sky. I moved the mirror so the reflection of the sun showed on the snow at my feet, rotated the reflection up away from me toward the Herc speck in the sky twenty miles away, rotated the mirror up and down to flash.

"We see your flash, Zero-Seven-Zero." *Hey, it worked, like in the movies.* "We'll come in low, make a pass, turn around for the drop," the pilot said. *Excellent.*

Soon, the Herc approached, buzzed over our camp low and loud, continued past us a few miles, banked, and turned around to make the drop.

"They're havin' fun with this," Dave said. The plane headed back toward us, slowed down, dropped lower, maybe two hundred feet above the snow. The Herc passed our camp almost overhead; we strained our necks to look up and follow it. A crewman stood in the open paratrooper door on the left side. He threw out a white bag. It plummeted down to the hard snow of the glacier.

"What the hell was that?" I asked the team.

"They just threw our mail out the door," Chris said.

"I can't believe this. No parachute? A white bag, on the snow, a mile away or more?" I said. "How can we possibly find it?"

"I'll go look," Steve said.

He mounted a snowmobile, guessed where the bag landed, and raced off in the direction of the drop. Twenty minutes later, he was back with the bag—to my amazement. We shared a big laugh of success, congratulated Steve once again on his sharp senses.

The bag looked heavy and lumpy. Steve put it on the snow, knelt, unzipped it, and reached inside. "It's gooey," he said, pulling out his hand, fingers, and a few letters covered in green mush.

"That looks like avocados," Chris said. Steve smiled, pulled out a note, along with remnants of pieces of avocados and apples crushed on impact at one hundred fifty miles per hour.

"Note is for you, Chris," he laughed.

Chris unfolded a paper smudged with green goo. "Hey, my pal in the Galley sent us some freshies from New Zealand, apples and avocados," she said.

We gushed into laughter—passed out letters from our friends and loved ones smeared in the just-made sauces. Steve handed me a few letters, from my mother, my friend Ken, a stamp collector, and two pink envelopes with a faint perfume air, my name and address written in perfect cursive—from Annie.

We read our letters, quiet, engrossed, later had fresh guacamole on our burgers, applesauce for dessert.

CHAPTER 21

Shadows in the Storm

Glittering white, shining blue, raven black, in the light of the sun,
the land looks like a fairy tale. Pinnacle after pinnacle,
peak after peak—crevassed, wild as any land on our globe,
it lies, unseen and untrodden.

—**ROALD AMUNDSEN**, *The South Pole*

Birchall Camp, December 27–29, 1989

I felt the cold trap me. I couldn't move. I struggled in my sleeping bag. I woke up. Blind, I gasped, pushed up my knit cap, pulled off my eyeshades.

Along my left side, the tent's wall weighted with packed, driven snow pinned down my shoulder. I pushed the canvas away with both hands. I sat up on my right elbow, looked across the food box at Dave, asleep. My watch showed 9 a.m. I read the barometric pressure on my watch, still low. I lay on my back, gloved hands on my chest.

Outside, the gale winds of Marie Byrd Land howled, buffeted the tent canvas, blasted it with driven snow, indifferent to our existence. *How can we matter?* The walls shuddered—cracked like rifle shots, snapped, cracked, snapped, the tent protested—relentless, endless. The wind

Dave in storm at Birchall camp.

carried sounds of subterranean echoes, earplugs little help. I looked up at the tent peak, a dim orange color, no sunlight to brighten it.

Each gust bent the two-inch thick hardwood tent poles. My mind drifted toward the idea that right now, this whole enterprise seemed crazy and my fault. Why should I, or anyone, endure this brutality? My fucking idea. Our circumstance felt irrational—the isolation, the risk. Six humans alone—no help or refuge for hundreds of miles. That fact sometimes terrorized me—more so when I lay idle in this biblical storm. *I've put all of us in danger. Because I wanted an adventure and bragging rights? Now I have one.* I felt stupid and selfish.

I thought of the world globe on my desk at home. I'd turned it upside down so that the South Pole sat at the top and I could see Antarctica. That told me the truth, the isolation of the Antarctic. Now we lay in our tents in one of the most remote parts of that continent. I imagined myself in touch with space, the white emptiness around us, and the blue-black sky above the blizzard—beyond the gale that trapped us in this void. The gale that repeated, *You are mine, you are mine.*

I considered the unthinkable—that the tent could rip away, expose us to the violence of the storm. Maybe kill us. Yesterday Dave and I had discussed that scenario. We lay stretched out in our bags and shared some hot Raro. We faced each other and talked while the wind moaned.

I worked up the nerve to ask a question with an obvious answer. I wanted to talk about the fury. "What do we do if our tent blows away?" I said.

"Put on our ECW, grab what we can, and get into any tent still up," Dave said. "If they're gone, grab shovels, dig a snow shelter, like in Survival School." I looked at him closely as he told me this. Looked for anxiety. He said this matter-of-fact. He made sense. I thought through the motions. Could we do that? Could I?

"I'm worried, Dave, scared in fact. Worried our tent won't hold up." I wasn't sure it was wise to admit being frightened.

"Yeah, I heard Chris is scared too. We're good." He took a slow swallow of Raro.

Now, Dave had been asleep over twenty hours, his way to deal with an endless storm. He could be an astronaut, sleeping his way across the solar system to Mars. While tent-bound, we talked, read, cooked, wrote in journals, and revised our field notes.

He had gotten engaged before we left. They planned to get married sometime soon, so we talked about that. They both owned homes. He wanted his fiancé to sell hers and buy into his. I'd been married twice, so I didn't have a great track record to give advice. That fact didn't stop me. I wanted to share what I learned from my mistakes.

"Why don't you both sell and get a new place together," I said. He liked that option.

Most of the storm, I had time to myself. Time to think. That usually led me to imagine what disaster could happen next. The other stressor was the burden of leadership as principal of this expedition. The rest of my team had their jobs, but they didn't have to worry about everything. That was my job, to be accountable. In a different place with a less hostile environment, I wouldn't have to sweat our survival on top of everything else. Antarctica had the upper hand all the time. I realized my good fortune that my teammates could balance me out—our mountaineers for ensuring safe passage; Steve with his extraordinary senses; Chris with her joyful spirit, energy, and down-to-earth practicality; and Dave, my opposite, always calm with few worries that I could see.

Recalling Christmas a few days ago led to thoughts of my son, of home and warm breezes. I wondered how he spent that day. I imagined us surfing; he would bring a friend. I was proud that he'd gotten good at that exhilarating sport.

I thought of my ersatz girlfriend, Annie. A relationship not very developed—maybe a few weeks into it before I left. I was clear in my head that I didn't dump my prior girlfriend after meeting Annie. I had a photo of Annie in a yellow-orange swimsuit, standing on the bow of a sailboat owned by one of her many admirers. She held on to the jib stay with one hand and waved over her head with the other. Outrageous. It made me smile to think of her. I had shown that picture to Steve. He threw his head back and laughed out loud.

I had to get the snow off my side of the tent to make room for me in my sleeping bag. My side faced northeast, where the storms came from. For cyclones in the southern hemisphere, with centers of low pressure offshore, winds rotated clockwise around the low. If you looked into the wind, the center of the low lay to your left, northwest in this case. We had set up our tent so the tunnel entrance faced at right angles to the wind, so snow didn't drift in, trap us inside. That meant my side faced northeast and took a lot of the beating. To get that snow off the tent, I had to go out into the blizzard.

I looked right, saw Dave lying on his back. "Hey, Dave—wake up."

He pulled up his dark blue knit cap that had covered his eyes, looked at his watch, and pressed the yellow barometer button with a finger in blue glove liners.

"Pressure's still down," he said.

"Yeah, good mornin'. Hey, I'm snowed in on this side; I need to move snow. How about a hand? We should check the tent skirt, make sure it's covered deep."

He groaned. "Yeah. We need the tent to be tight. Okay. Guys are comin' over for brunch in a while," he added. "After some breakfast and a piss."

Done with a small oatmeal breakfast, we suited up. That meant all our ECW including the Big Reds, which we put on while inside our tent. Boots we retrieved from inside the tent liner on the right side. We put them on last and untied the tent tunnel. Dave crawled out headfirst. On my hands and knees, I followed his wide, black butt.

Outside I saw frozen violence. Snow blew sideways, gray and white, no definition. I looked in the direction of the horizon, hopeful I could make out something. No. I looked up, searched for a piece of sky. No. The wind caught me, pushed me around. I got its message. *Get out of my way, you puny nothing.* With snow up to my knees upon each step, I struggled to walk. Trail flags set up in camp to mark pathways stood horizontal, pointed southwest where the wind headed fast as a freight train. The flags snapped in the wind; edges now frayed. I saw the radio antenna still strung up—good, but so what? We wouldn't call anyone. Snow snuck into gaps in my parka, then melted; I felt cold drips down my neck. My hood tunnel extended to shield my face, muffling sounds, but the moan and hiss of the gale dug into my head.

On my side of the tent, each with a shovel, we moved heavy snow off the tent wall, careful not to rip the fabric. We piled snow up high on the tent skirt. Did the same for the other sides. How long would this touch-up last? My goggles crusted over with snow every few minutes. I brushed them off with my gloves, kept my shovel in action. We shouted at each other, decided that we'd done what we could, found the tent tunnel, untied it, and crawled back inside.

Relieved to escape the wind, we brushed off caked-on snow in the tiny space at the entrance, one at a time, careful not to spill snow on the sleeping bags. I took a whisk broom and cleaned snow off Dave, then he returned the favor. I pushed the invasive snow pile down the pee hole by the entrance. I then realized I had been cold outside, very cold. My face had numb spots.

"Good as it can be," I said. Dave gave me a nod; melting snow on his cheeks and beard beaded and dripped.

An hour or so later, we heard loud, urgent voices outside over the gale: "We're all here. Untie!" Sounded like Tucker. Then Chris, Steve,

Cain, and Tucker crawled inside fast; each in turn pulled off their Big Red and stuffed their boots inside the space between the tent and tent liner. We kept the new snow under control, brushed up after each person. I looked at their faces, burned by the sun and wind, with pale eye sockets. They didn't act agitated or fearful. Lot of laughs and lame jokes about the storm. I thought about adjusting my attitude. I tried the calmness on for size. It fit.

Tucker had brought his rucksack.

"I've got brunch," he said. "Pancakes and dehydrated peaches." He began to hum a tune, unloaded his pack with the goods.

On our small one-burner stove, Tucker made pancakes for all. The tent warmed up from our bodies and the tiny stove. We sat on the tent floor, separated from the snow by only a few inches—a tarp, insulated pad, sheepskins, and sleeping bags. The six of us squeezed in, rested on or against parkas and sleeping bags.

Tucker did the cooking steps with élan. He had fun. His hair a wild dark spray, he wore his red-and-black wool plaid USAP shirt. His beard thicker than weeks ago, his wire-frame glasses a bit bent from abuse.

Tucker cooked one pancake at a time in a small pan on the small stove, melted frozen butter, passed out the food. I learned pancakes have a reassuring, home-cooked aroma. Chris got the first one. "Here you go, Chris. Ladies first," he said. Spirits perked up, even mine; I felt reassured. We tried the resurrected peaches. "Not as bad as I thought," I said. All nodded, appreciating this minor luxury. Talk went to the storm, how our tents had held up.

"My side of the tent got buried, so Dave and I went out to move snow. Nasty," I said.

"That's way better than snow blowing off the skirt," Tucker said.

"Yeah. That's how tents blow away. Wind gets under the skirt and goodbye," Cain said.

"Is this storm worse than the first one after Put-In?" I asked everyone.

"Sure is," Tucker said as he cleaned up his pan.

"Oh yeah," said Chris, followed by murmurs of agreement from Dave, Steve, and Cain.

We compared barometer readings, decided we all had the same unpleasant story. No change—pressure way down.

"I went out last night to take the weather, wind speed, and direction," Steve said.

"That sounds like a shitty experience. Glad you volunteered." I was gladder nobody had expected me to do it.

"Gusts up to forty knots, sustained at twenty-five," he said.

"You called that to Mac Center, they say anything?" I said.

"Thanks, I think," Steve laughed. We all joined him—nerves?

"Yeah, they couldn't care less," I said. "What's that make the wind chill? Wait, I'll look that up." I got out the table in my notebook. "Hey, it's about twenty degrees and wind forty-eight miles per hour, wind chill has to be about minus twenty-three Fahrenheit."

"Ouch," Tucker said. That brought more chuckles. "Is that Condition Three or Condition Two in McMurdo?"

"That's not even Condition Two, more like high side of Three, 'normal weather,'" I said. More guffaws.

"Good to know that," said Tucker with a sideways grin.

Well, we would live in warm buildings there, the warm Galley, warm bunks, not the same as a tent in a storm and this emptiness. But it was ugly in Mac Town—not so here.

We ate; changed positions every few minutes to wake up numb feet, legs, butts; chatted about nothing. Not about our tents blowing away. Spirits sounded high. I was happy to learn that we were solid, that we could handle this.

I stirred the peaches in my mug, took a scoop to my mouth, realized survival is about how to cope with a little hope added. "Well, guys, I guess no beach picnic today," I said.

Brunch over, the guests gone, Dave and I straightened up.

"I'm gonna shave my beard," Dave announced.

"Why? It's sun and wind protection," I said.

"Something to do, a change." He dug out his razor, melted a block of snow into a pot of water.

Bruce digging out his tent after storm.

In a few minutes, his face looked like a map with snowy peaks and red valleys. Oh well. Sunscreen was a must sometime soon.

I got out my book, rolled open my sleeping bag, lay my head and shoulders on top with my Big Red for a pillow. My earplugs in to smother the wind blasts, with gloves and hat on, I read.

Before we left, Anne, a woman who had made an expedition deep into MBL a few years earlier, had given me advice. "Have you read *War and Peace*?" she asked and laughed when she heard of our destination in the Ford Ranges. "That novel is about the right length for a seven-day storm." I'd made no response but had that oh-shit feeling. I didn't bring that novel. I had *Lonesome Dove* by Larry McMurtry and assorted other books including *Discovery* by Richard E. Byrd, about his Second Antarctic Expedition in 1934–1936, the one when he'd explored a new land that he'd named after his wife, Marie in 1929. I checked the book out from the UCSB Library and shipped it south to Antarctica. Didn't get permission to do that.

I began to read Admiral Byrd's book in the evening and in stretches of time spent tent-bound during storms and whiteouts. Byrd had written a firsthand account of their discovery of much of western MBL and of the first men who traveled by dog sledge to reach the edges of the mountains where we now camped.

I decided it would be fun to inscribe the inside cover of this library book. I wrote, "Read in my tent in Antarctica during the storms of Marie Byrd Land," then signed and dated it.

Years later, a student in one of my classes confronted me with that library copy of *Discovery*. She opened the book, pointed to show me what I wrote. She smiled; I laughed, but I remembered the nightmare.

CHAPTER 22
||||||||||||||||||||||||||||||||||||||||||||

Down the Ochs Glacier

*So we arrived and were able to plant our flag at the
geographical South Pole. God be thanked!*

—**ROALD AMUNDSEN**, *The South Pole*

Depot Camp, December 30–31, 1989

"We marked a route, set flags all the way down the Ochs," Tucker said. Ochs Glacier was our downhill path to the massive Balchen Glacier where we planned to go next. The Ochs is a tributary that joins with the Balchen Glacier. The Ochs runs down to the north. The Balchen trends east-west between the Fosdick and Phillips Mountains and flows west down to Block Bay. From the Balchen, we would access the migmatites on the north cliffs of the Fosdick Mountains.

"Set the flags at two-tenths-mile spacing on the east side," Cain said.

Made me glad to hear they planted flags down the Ochs and at two-tenths spacing.

"Great news," I said. "Any crevasses?"

"None were obvious, but there're some stretches of blue ice on steeper slopes."

"We had to drill holes to set flags in the ice," Tucker said. That sounded like a warning to me.

Cain, Dave, and rock samples at Depot campsite.

"We can get down those slopes, right? No one can stand up on a blue-ice slope," I said. The result of virtually all the air space driven out of snow and ice by time and pressure, blue ice was slicker than a skating rink. You could see your face reflected in it.

"Should be fine; we'll set up rope brakes for the sledges. We've got studs on the snowmobile treads," Tucker said.

"Amundsen used rope brakes," Cain said. "Won't travel in two trains like we did to Birchall. Each snowmobile tows one sledge or one sledge and handlebar sledge at rear for braking."

"How about two snowmobiles and a sledge in between tied to them?" I asked. "Lots more traction and control."

"That will take three trips. Each trip adds time."

"Yes, I know. We've got time to do this."

"More trips mean more chances for trouble," Tucker said.

"Okay." I paused. "Makes sense."

But the fact that Amundsen did this, a man who didn't break into a sweat reaching the Pole, didn't mean the whole expedition would be easy—no choice; we had to get down somehow.

* * *

We had set flags at half-mile spacings on the route to Birchall Camp the last week before Christmas. That proved too far apart in low visibility conditions. I thought back to our Christmas week experience at Birchall and what we'd learned about traveling in tough conditions. After the three-day blizzard ended at Birchall, fog came in. No miserable wind and snow, but we couldn't see to make our way back to Depot Camp on the Chester snowfield. A mental wreck, I worried that our supplies would run too low and we would be forced to grope our way to the Depot with no chance to find it.

How long will the fog hang around the Birchalls? I'd wondered. The storm and the fog cost us time and used supplies. We had maybe two weeks left. I spoke to Tucker. We stood on the snow in the fog.

"This whiteout could last days; I'm concerned about our supplies. We need to get back to the Depot soon," I said to him.

"We should be okay. Plenty of food, still on our frozen supply, and we have dehydrated backup," he told me.

"What about stove fuel?" I worried that we could run low of crucial fuel—or even run out. Then we'd have no water.

"Using about a gallon and a half of fuel per week per tent, I figure we have nine days of reserve," he said.

"That's good, but we can't sit here nine days anyway. That wastes a week of work. Do you think we can get to higher ground, above the fog?"

"The trail flags are probably all blown down."

"But we made trail at the break of slope, foot of the peaks. We can find that in the fog and follow it."

"Maybe, but if the fog is on the Chester snowfield, up high, we won't find the Depot," he said. He was right. We couldn't take the chance of getting lost there. We needed to wait it out.

Then, late Saturday near midnight, Steve stuck his head into Dave's and my tent. "Fog's lifted, pulled back," he said without comment. I joined him outside, saw clear sky. I looked north and saw the edge of fog down low in Block Bay.

I went over to Tucker and Cain's tent.

"Hey, we can get out. Should we go now or after we sleep some?" I asked them. They replied that we should rest up. "How about we put in new flags on the route up to the Chester snowfield—at two-tenths mile?" I said. "If the fog comes back tomorrow, we can find our way above it to higher ground on the Chester snowfield."

Cain and Tucker thought this was a good idea. Steve, Tucker, and I got a bunch of flags, drove off, and retraced our route, planting flags every two-tenths mile. Close enough together to see them if fog came back. Most of the flags we'd set on the way down to Birchall had blown over. We reset them. We finished at 2:30 a.m.

I turned in for some sleep and woke up to find a bright clear blue sky—excellent day for travel. Relief. We made it back to the Depot in a few hours, but Tucker's snowmobile broke down and needed to be towed—always a wrinkle in any plan.

At the Depot, we made camp, Cain and Tucker flagged the Ochs Glacier route, and we slept overnight. We planned to work east and up the Balchen Glacier along the front of the Fosdick range. Our first camp would be at the foot of Mount Avers.

Besides discovering the metamorphic minerals at Birchall Peaks that indicated the peak heat and pressure the Fosdick rocks had undergone, Chris and Steve had found evidence of multiple episodes of deformation; the rocks had been squeezed or stretched at several different times from several different directions. That finding needed to be checked along the range to see if the deformation sequence was consistent through all the Fosdick Mountains. If so, it would suggest a larger-scale plate tectonic cause.

* * *

The next morning, New Year's Eve, we prepared for our trip down the Ochs Glacier to the Balchen. Dave and I packed our sledge. I lugged the gear from inside our tent onto the sledge, no spring in my step.

"Hey, Bruce, you okay? You're fumbling around, moving slow."

"I'm tired and in pain, Dave." He didn't respond.

My tendonitis had flared up, I figured from breaking and making camp three times in thirty-six hours, and from digging out our supplies buried in the blizzard. I moved slowly, fuzzy with exhaustion, forgot what to load, looked for stuff I knew we didn't want. Everyone else acted fine, no complaints—well, they were ten and twenty years younger than me.

Our sledge sat on the snow with its load piled chest-high. We threw the green canvas tank over our gear. Dave on the opposite side, we pulled it down snug, then lashed the canvas to the sledge.

Feeling an urge, I took off my gloves, tucked them under my armpit, unzipped my fly, and dug in my pants through layers of underwear to take a piss. I watched the stream melt a hole in the snow. Done, I stuffed myself back in my pants. Dave looked at the snow.

"Hey, look," he pointed to the stained hole in the snow.

"What?"

"Your piss is brown."

"What the hell?"

"Man, you're dehydrated, Bruce. Piss should be clear. Get some water in you now. We gotta get moving; you need to be sharp going down the Ochs." Dave left to get his rucksack.

I grabbed my thermos out of my pack and downed it over a few minutes. I had less than an hour to recover before we left. How could I make such a greenhorn mistake? Lucky that Dave noticed. When you're cold all the time, you forget you're thirsty; you forget to drink. I had to monitor my fluids, urine color, have discipline.

Packed up, recovered, I stood still and surveyed our train of snow-mobiles and sledges. The sky clear, with a few clouds drifting, I faced the morning sun in the east. Sunlight reflected off new snow and caused it to sparkle, like ice jewels in a cotton blanket. I took in a deep breath of good pure air and sensed it flow into me.

We drove off to the north. I felt back in form, my head clear but tendon still hurting. I towed one sledge, Dave another, Cain and Tucker two sledges each, with Chris and Steve at the rear of the last sledges as brakemen.

An hour later, we arrived at the top of the Ochs, the east branch, and stopped. Four snowmobiles lined up abreast of each other, facing down the glacier. I had not seen this side before. I looked down it—saw Block Bay. It held the ice that flowed off the Balchen Glacier from the east. Across it, the Phillips Mountains lay under thin clouds about fifteen miles in the distance. The Ochs looked steep from where we stopped.

I remembered from the recon map that the Ochs was about four miles long and dropped almost seventeen hundred feet to Block Bay. Steeper sections of the glacier showed blue glare ice. I could see the flags Cain and Tucker planted fade into the distance. We'd never had to go down steep slopes like this before nor had any of us traveled on

Travel on the Balchen Glacier.

blue ice. Our sledges weighed hundreds of pounds, cannonballs sliding down the slope.

Today could turn to shit.

I kept quiet, decided to trust Cain and Tucker. "Let's get rope brakes under the runners of the sledges Bruce and Dave are towing," Cain said. We unloaded two lengths of thick rope about two inches in diameter. I lashed one end of the rope to a side of the sledge frame. I lay down on the snow, dug a slot with my hand under the runner, passed the rope to the other side, dug another slot under the opposite runner, and tied that rope end to the other side of the sledge. My sledge sat on its runners and the runners on the rope. We rigged Dave's sledge the same.

"Don't follow each other downhill. Spread out in case someone loses control," Cain said. "Go straight down; don't traverse or the sledges will get away from you." We began to head down, stayed near the flags, abreast of each other, about twenty or thirty yards apart.

Upper parts of the slope looked snow-covered and proved okay to drive. I checked my rope brake. It'd come off one side of the sledge. I stopped. The rest of the team went on. I lay down beside the runners to shove the rope back under the sledge and retie it. Rope attached and under the runners, I started down once again.

From the top edge of a steep section, I could see Cain, Tucker, and Dave driving downward. I crossed a snow-covered slope with a few blue ice patches. The rope brake worked to slow my sledge. I crossed a gentler stretch of the glacier then came to another steep part. The team made it down ahead of me. They looked up the glacier and watched. I drove the last stretch.

This section appeared even steeper than the first part. I drove onto blue ice. I gained speed. Turned to look back at the sledge I towed. The rope between my snowmobile and the sledge drooped slack, the rope brake now untied and loose again. The sledge slid faster than I drove. I needed to pull the sledge so it wouldn't overtake me, whip me around out of control—I'd cartwheel down the slope. *I won't let that happen.* I tightened my grip on the handlebars, pressed my knees against the seat between them, applied full throttle, sped up. The rope jerked tight, then went loose. I looked back and saw the sledge off to my right-rear side. The sledge ran free, about to overtake me.

Up ahead, I could see the team looking at me. The slope evened out. At full throttle, the rope pulled taut again. I couldn't go faster; I sped past the group. The glacier surface almost level there, the snow slowed me. I stopped my snowmobile; the sledge obeyed, stopped too. Fury rose inside me.

I shut down my machine, got off, marched over to Cain. He stood next to his snowmobile, rotating his arms in big wide circles around each shoulder, one and then the other, to move blood to his hands to warm them.

"Cain, goddam it, did you see what happened? I almost got whipped outta control. I told you we should've done two snowmobiles on a sledge." Cain stared at me through his goggles, without expression. He looked away, waved the others to start up and follow him. He turned back to me, squared his shoulders, gave me his shut-the-fuck-up look, got back on his machine, then headed off.

I joined them, followed, drove on, and headed to the base of Mount Avers. That was my mistake to blame him. Who knew what to expect? The others made it down fine. I lost control; the sledge didn't. Forget about it. KIT.

CHAPTER 23

Avers Camp

...Soon we came to sledge tracks...
Amundsen had beaten us to the Pole.

—**LAWRENCE E. G. (TITUS) OATES,**
Robert Scott Polar Party Member, 1911
(attributed by Beryl Bainbridge in *The Birthday Boys*)

Avers Camp, December 31, 1989–January 3, 1990

"Hey, what time is it?" I raised my voice over Eurythmics' "Would I Lie to You?" The song blasted from my yellow speakers suspended from the clothesline. Cold hadn't killed the batteries yet—they would be dead soon, though, and I had no others.

"Who has the right time?" Dave asked. "It's important. We need exact seconds, a countdown." "Yeah," a chorus agreed. None of us had the exact time—we all knew that. The sun was low in the south behind Mount Avers, near midnight, a rough clue. The right time would be New Zealand time, used by the US in this part of Antarctica.

"I've got two minutes until," I said.

"My watch says it's seven," said Chris.

"My watch is ten seconds faster than Bruce's," said Tucker. "Let's call somebody, get the time."

Travel at night on Balchen Glacier.

"Are we going to sing something at New Year's?" Steve asked. "Who knows the words to 'Auld Lang Syne'?" Nobody did.

"It's a Scots song," said Cain. "Dunno all the words though."

"I can hum it," Chris said. Everyone said they could start the song. We all sang the opening words, "*Should old acquaintance be forgot...,*" then began to hum, la, la, la, but didn't get too far.

Tucker switched on the radio, turned on its speaker. I turned off the music.

"Byrd Surface Camp, Byrd Surface Camp, Sierra Zero-Seven-Zero," he called them, our nearest neighbor four hundred miles east. No answer. Static whistles, whoops. He repeated the call, waited half a minute, then ended with "Happy New Year anyway, we can't copy."

"We can't be heard from our campsite, too close to the cliffs to transmit," I said, "but we can hear others." *Hope we don't need to call in a Mayday here.*

Chris pulled out a couple bottles of champagne. We had brought them from McMurdo—kept them buried in a small snow pit with frozen foods. Hadn't gotten cold enough to freeze them and blow the corks.

Chris held a bottle. "I'm gonna get this ready to pour, should crack this sucker at midnight—close enough," she said. "Get your mugs out." She popped the cork. Cheers sounded.

We compared the time we each had again. We couldn't decide on the correct time. We never thought to synchronize our watches.

"What else could we sing?" Tucker asked.

"Something from *Oklahoma!*," said Steve. Laughs around the tent, hip to hip, shoulder to shoulder, legs up against others, feet searching for vacant space.

The radio came alive: "Byrrrd Surface Camp, Byrrrd Surface Camp, Mac Center." We stopped our chatter, listened. Mac Center repeated the call; Byrd Surface Camp didn't answer. Heavy static—the voice from Mac Center faded in and out.

"Mac Center, Mac Center, Byrrrd Surface Camp. Happy New Year, Beau—I have the one-two Zulu weather." The Byrd camp radio operator called in the observations to Mac Center.

"Hey, twelve hundred Zulu, it's New Year, 1990, a new decade," I said. Zulu time used for weather data was twelve hours behind New Zealand time, so it was officially after midnight in MBL. Whoops and cheers sounded in the tent—Happy New Year at the loneliest place on Earth.

Chris poured champagne all around. "I've got noisemakers," I said. Dug into my pack, pulled out paper cone-shaped party noisemakers—kazoos, horns, and coiled paper squawkers to blow and aim at each other—and passed them around. Noise erupted.

"Bruce knows how to put on a New Year's party, I'll tell ya," Chris said. More noise.

"Hey, key the mike, key the mike," I said to Tucker. "Share the joy." He pressed the microphone key; we made a racket of joy for part of a minute. Tucker released the key, waited, but nobody responded to our greetings.

We blew our noisemakers, took another drink of champagne, and blew them again. At a pause, we heard other stations calling in wishes to each other, none to us.

"Everybody, toast," said Chris. Mugs clinked.

"To 1990," I said, a sensible toast.

"Up yer kilt," Cain shouted. Yeah. That was more like it. Clinks.

"Anything worth doin' is worth doin' to excess!" Steve said. Yeah. Laughs and shouts. More clinks, more noisemakers, no reason to stop.

"Wherever you go, there you are," Tucker said, raised his mug.

"Yeah, Buckaroo Banzai," I said. Clink of cups. Cheers. Fizzy sweetness poured down my throat. "Dave and I were just talking about Buckaroo today. Why do we keep moving camp if we're already there?" I asked. Laughs, clinks, and cheers followed.

"Skido needs a drink," Tucker said. He pulled his mouse mascot out of his pack and stuck its long pointed nose down the neck of the champagne bottle. "Yeah!" Everyone approved.

"'Tis New Year's. I usually dance a pas-d'-bas; it's Scottish, you know," Cain said. He held his now-empty mug. He felt pretty good right now, I guessed.

"Yeah, let's have a jig," Chris said, raised her mug. Others chimed in—yeah, a pas-d'-bas. Not sure about that dance—ballet, or something foreign, radical? That would be weird, maybe fun. A chant started: "Pas-d'-bas, pas-d'-bas." I joined in anyway.

"Dun have music. Can't without it," he said.

"Give us the beat. We'll clap—don't need music," Tucker said to encourage him.

"Come on, don't be a weenie," Chris said loud with laughter.

"No space fur it," Cain said.

"On the pots box," said Tucker. Yeah, everyone agreed. Hop up on the pots box. Come on!

"Okay, standby," he said. He climbed atop the wooden lid of the box in the center of the tent—about eighteen inches above the snow level. Standing up, his head didn't reach the tent peak at eight feet. He stood inside the perimeter of the clothesline that ran square around the interior of the orange tent peak and framed his chest. It held damp socks and glove liners attached by clothespins. A multicolored, foldable Happy New Year sign stuck out behind him. His boot's blue canvas FDX issue were calf high with dark blue leather toes and white laces, clumsily over-sized for insulation but with good black cleat soles. They weren't ballet slippers. He wore his red-and-black Scot plaid wool shirt and many-pocketed black wind pants with black suspenders.

"It's like this," he said and began to clap in four-four time. We joined in.

He clumped his boots on the box lid and held his arms by his side, hunched over a bit to fit in the peak of the tent. Hop on right boot, slide left boot, hop on that one, hop on right, kick left boot, then hop on that boot, switch feet, repeat. The wooden box drummed as his boots hit. *Clap-clap-clap-clap*, we cheered him on. Claps and boots sounded in unison, along with shouts of encouragement. This looked authentic, I admitted, didn't know otherwise.

He came to a stop, jumped down to a small empty floor space between all our feet, and accepted our congratulations, cheers, applause, and laughter. He showed a rare grin, wide. Surprised, I laughed, held a smile. I had just learned Cain could be fun.

January 1, 1990

I crawled out through our tent tunnel to see a calm, blue morning sky of the New Year. I stood up, looked north, felt the sun on my face, let it soothe me. Happy in the almost-warm air, I surveyed the scene about us. The rock wall of Mount Avers stood straight up behind our camp, gray and free of snow. *That's near vertical*, I realized. The map said the top reached about two thousand feet above us. Couldn't tell that by eye with nothing for scale.

Behind our orange tents, a jet-black volcanic cinder cone lay at the base of Avers' cliff face. Humped like the back of a basalt whale the size of a city block three stories high, snow lapped at its edges, mimicked a frothy sea. No visible signs that the movement of ice had disturbed the cone in the way it had clipped away the rock on the north side of Avers. The cinder cone had to be younger than the last high stand of ice thou-sands of years ago. That cone could have been created yesterday. The idea caused me to focus more.

These mountains weren't in a region where volcanoes are common, like a plate boundary. Mount Erebus, still active, was almost a thousand miles away. Other volcanoes that could be active, but not known because they couldn't be monitored, lay hundreds of miles to the northeast where the WAVE team was working. Not only that, but Mount Perkins, a volcano down the Fosdick range about twenty miles east of us, also could be geologically young; early studies by others suggested a couple of million years. Perkins had been dissected by glacier movement, so this cone at Avers was younger than Perkins. I wondered about these facts and what would explain why the region where we found ourselves would have had volcanic activity. My wonder expanded.

How improbable to stand here at the bottom of the world in this wilderness—embraced by the solitude, the infinite, by the white, the blue, the peril, and the impossible magnificence. No clouds or wind at noon, our shadows lay short and blue on the snow. Most days, the sky held clouds that either blocked the sun or traversed the sky at speeds that depended on what force of nature was to come next—a blizzard, a whiteout. The wind could be seen when it blew snow horizontally across the frozen air. I'd never forget that last storm. Four days of a cruel blizzard and whiteout kept all of us trapped inside our tents. I felt a familiar emotion—days of struggle replaced with seductive, sublime relief.

The Balchen Glacier adjacent to our camp lay flat like a lake surface, a lazy look for this immense glacier that had drained part of the West Antarctic Ice Sheet. I could make out the edge of that sheet up the glacier thirty miles or more to the east, on my right. The ice thickened in that direction and smothered any mountains that might've had the misfortune to lay in its way. That ice edge promised a terrain beyond my sight that stretched flat and empty for thousands of square miles. What would I feel if I found myself inside that emptiness—alone? Would I go insane? The ice sheet drained through the glacier on its way to Block Bay, and the cold, black Southern Ocean sixty miles to the west. The glacier looked quiet and tranquil, its movement almost a secret, its cargo not obvious. I thought about how much ice ground its way past the spot where I stood. The scale overwhelmed me. I felt dazed.

No one up yet, my mind wandered to other topics, such as my own magnetics research program. My grad student, Chris, functioned at high efficiency but focused on the metamorphic rocks. She had the expertise on that topic, not me—but I needed to find a way for our research to intersect so I could supervise her graduate work. I knew that she and Steve took observations on deformation features in the rock outcrops, trying to find evidence of the directions of stretching and its timing. Deformation that accompanied the formation of the mountains probably occurred a hundred million years ago, at slightly different times and places. Finding coherence in the deformation was important. Figuring this out was tough stuff but would allow us to unravel the history of the mountains and the fragmentation of Gondwana. I had not anticipated the significance of this part of her work.

My mind negotiated with itself. It wasn't reasonable for her to take on some of the paleomagnetic work; that was my field. But the deformation patterns of the Fosdick range—that might make sense. She could work up rock magnetic data, the AMS, on my core samples in the lab. That data would reveal directions of stretching the rocks endured. Some magnetic minerals in rocks are elongated, and when the rock is stretched, these minerals line up in the direction of deformation. That direction can be measured in the lab. The cores I drilled were carefully oriented. The lab measurement is quick and many cores could be analyzed. They would give more accurate directions than structures found in the outcrops where alignment features were rare and difficult to measure. That was a win-win. Otherwise, I would have to bring in another grad student to cover some of the magnetics program or replace her. But that was a non-starter. I couldn't do without her energy and experience in the field next time we came out here. She was one of the best graduate students I'd had—a self-starter and expert in her work. Her skill level had impressed me. Her work would make a big impact on our project. Besides, replacing her would be a chickenshit move by me, even unethical.

Just about then, I saw Chris and Steve emerge from their tent. They had mugs in their hands. Chris had on her usual for a fine day: blue wind jacket and pants, pink balaclava, glacier glasses. They took in the morning, looked over to me. Chris walked over to say hello. Was now a good time to talk about this? Why not? We exchanged greetings, small talk.

"Chris, I've been thinking about how to get you involved in the magnetics program," I said. "We haven't talked about this, but as my student we'll have to carve out some work that makes sense for our collaboration." She lowered her mug, looked at me, silent, eyes hidden behind her black glasses.

"I think that AMS will work to your advantage, for Steve too," I said. "You know—the anisotropy of magnetic susceptibility," I added, to make sure I was clear. "It could get you accurate deformation data from the oriented cores I've drilled. I need you to take charge of that part." I felt intensity rise in me, convinced I had come up with a great idea.

"You never told me that before. It would change my work plan," she said. I sensed a rise of emotion in her too. "I'm in way over my head now with the Fosdick rocks," she said. "I can't learn AMS, don't know a thing about it."

"The AMS will help you; the data are better than visual estimates, and you will get more data," I said, more intense now. Both of us were pressurized, but I couldn't stop. Chris looked at me, paused.

She began to cry, turned, and walked away. I stood on the snow, stunned—had not expected a meltdown. She went over to Steve, had a short talk with him, and ducked into their tent. Steve walked over to me.

"Hey, Bruce, what's up with your AMS plan for Chris?" He appeared calm; he gazed into the distance.

"We'll get better data," I explained. "It's a hundred times faster than outcrop measurements and ten times more accurate. It's a way for me to get help on my program and for Chris and me to collaborate." Steve paused to think; he knew about AMS, knew its value in determining past directions of rock deformation.

"Chris feels it's too much for her to handle. She feels you're changing the program in the middle, that it's not fair," he said.

"Maybe I am, but it will work out better for all," I said. Steve thought for a few seconds.

"I'll see if I can get her to buy into it," he said. He walked away to join her in the tent.

Now, as I stood on the snow in this beautiful spot, I didn't feel so happy. I had blundered through that encounter with Chris, had been too pushy and maybe squandered a good chance to work with her. Bad timing? That happens when I don't stop and think before I speak. I'd blown it. I needed to not act like an asshole, not to Cain, and never to Chris.

CHAPTER 24

Bird Bluff

There is no escape anywhere. You are hemmed in on every side by your own inadequacies and the crowding pressures of your associates.

—**RICHARD BYRD**, *Alone*

Bird Bluff Camp, January 3-4, 1990

The night after we arrived at Bird Bluff, Chris demanded a science team meeting to challenge my work plan for her to do the magnetic measurements—a source of anger for her. This would be the first time we spoke since our confrontation at Avers Camp four days ago. Dave and I sat on my sleeping bag, facing across our tent at Steve and Chris.

She began, tension in her voice, defiance in her body. She asked me to defend my reason for the magnetic work: What is the value? Why didn't I bring this up before? Why didn't I suggest that this work be part of her thesis prospectus? She wanted witnesses to this conversation and to my answers. Steve acted less aggressive. He saw the reason for the AMS but felt uncomfortable with a new work agenda put on the table. Dave thought it made sense to do the AMS; it added value to our program.

Phillips Mountains across Balchen Glacier from Bird Bluff.
(Photo: Steve Tucker)

I felt chagrined at not having anticipated the need for that research and for how I handled Chris, my conflict with her. Even though her attitude felt out of line, I knew I had provoked her and needed to give some ground.

Chris made a compromise proposal: I do preliminary work, and if it showed value, she would complete it. First time I could recall one of my grad students suggesting work for me. Her idea struck me as impertinent, but I suppressed my reaction.

"No, that won't work for me; this is your project to do," I told her. Steve spoke up before she could respond, said he would do preliminary work and then we could evaluate where to go based on that. We all agreed. Lesson learned? Don't throw my weight around. That doesn't get results, for sure not with Chris.

January 5, 1990

We had arrived at Bird Bluff night before last, worked our way east from Avers Camp, stopped to sample and map at Mount Bitgood along the way. The recent storm had pushed new snow to the south side of the Balchen Glacier up against the Fosdick range front. Travel went smooth and quiet over the softness of the new carpet, the weather calm and clear. Yesterday, five of us had traveled up Reece Pass glacier south of camp to survey and sample Mount Richardson. Not me. I stayed behind because of the tendon pain I had been dealing with. We had to bend the rules. Two mountaineers with the team would be safer for travel on a glacier.

That morning, Cain and I had gathered our gear for a trip to Bird Bluff while the others, led by Tucker, prepared to travel twenty miles to the east and back in one day. A long trip. Forty miles. They planned to visit O'Connor and Griffith Nunataks to map and sample those rocks at the very far end of the range.

"My groin is killing me," I had said to Tucker. I shifted my weight to favor my left side. "I don't think I can make it forty miles, wouldn't be smart. I can work here at Bird Bluff. But I don't feel good bailing on this trip."

"Bummer. Okay, you're the boss," Tucker said. He looked at Dave.

"We need to survey the far-east end of the Fosdicks. I want to go, but I can't," I said to Dave.

"You're gonna miss out, Bruce."

"That would be a hard trip. It hurts me a lot to drive a snowmobile." But then I felt worse. He was right.

Tucker said, "This could work out. Cain and I figured we don't have enough fuel for all of us to make it forty miles anyway; two snowmobiles must stay behind." I knew Tucker wanted to make this trip.

"I'll stay with Bruce," Cain said. He didn't argue with Tucker. Maybe he didn't want a full day in the saddle?

"Hey, thanks. I want to drill rock cores on the face of Bird Bluff—the cliff," I said.

"What do you know about Bird Bluff?" Cain asked.

"I've studied the air photos. There's a deep wind scoop in front, maybe a hundred feet. I remember it from the Recce flight. We could start at the scoop bottom and work our way up the sides, cover a lot of the cliff face."

"So, you want to start at the bottom, eh? Think you can climb down and out with your pain?"

"I think I'll be okay."

"Why this scoop?"

"It's the deepest scoop along the range, steep front slopes, bottom's flat, runs up against the cliff face. Great exposure of Fosdick rocks," I said. "The cliff top is maybe three hundred feet above the scoop bottom, so we could get access to a third of it."

"Sounds sensible if we can get down into it. Where're the birds?"

"When the bluff got named, there must've been birds here—snow petrels. They're farther down the glacier now. Near Marujupu, remember?"

Tucker and Dave got on their machines and headed off east. Steve and Chris stood on the handlebar sledges towed behind. It felt shitty to see them go. I wanted to experience the end of the range, where the last solid rock could be stood upon, at the edge of the impossible ice sheet that buried everything that could be imagined about the ancient body of Antarctica. But because of my pain, I was glad I didn't have to suck it up today.

Next to my snowmobile, loaded for our day of work, I looked south to Bird Bluff, a gray cliff that wore a frosting of ice at the top. It presented a gray scarp of metamorphic rocks with buff layers of granite. From the map we had, I couldn't tell the distance to the bluff because I didn't know exactly where we were camped—maybe a few hundred yards; it appeared that far by eye, but in the Antarctic wilderness, with nothing for scale, it proved impossible to estimate distances. The air appeared so clean and clear that no landscape faded from view in the distance; all features stood out in detail.

Cain loaded his snowmobile, continued to check his food, water, ropes—I paused, thought about the quality of the air that surrounded us without corrupt particles of civilization. I took in a slow, deep breath of the silence, appreciated the uniqueness of the virgin atmosphere. I feasted on molecules that had not entered a human until now. I looked up to survey the peaceful West Antarctic sky above me.

I adjusted my pack on the snowmobile, then faced north to our three pyramid tents planted on level snow. Behind them, the width of the Balchen Glacier spread for ten miles to the base of the Phillips Mountains, gray-blue in the distance, low in profile but sharp in unexpected detail in the dry air. The sun had moved to the rear of these mountains, painting indigo shadows on the slopes that faced me. If I stared at this view too long, the rare visual impact would capture me, hypnotize me.

Like every panorama I witnessed over the past weeks, this one appeared unfamiliar. Not because I hadn't seen one like it before. The light was always present, but its texture changed quickly to make each scene transient and unique. The sky, rock, snow, and ice made up the only distinctive variables in this unfamiliar view. Absence felt acute: no vegetation; no leaves colored for fall; no tracks of wildlife puncturing the snow; no roads made by humans slicing across the ranges; no contrails in the sky; no wisps of chimney smoke that suggested hidden cozy cabins; no sounds of traffic; no sign of life as I knew it. I realized the source of my inspiration for this place—nothing was familiar in any sense—its beauty peculiar, its views unexpected. Disbelief erupted moment to moment. Splendor. The landscape held surprises, including me.

My daydream was interrupted by recalling the work ahead. I dug into my pack on the machine. I inspected my rock drill, made sure it was ready for today. All checked out, I looked up at the Fosdick Mountains that bordered the Balchen to the south. The mountains stuck up from the glacier in places to form gray cliffs which reached more than a thousand feet above the snow. Down the Balchen to the west, the range cliff tops looked like castle parapets. I expected to see shadows of archers along the edge. In front of us, ice had dissected the Fosdicks into ridges, punished the mountains in the past as it moved from the range tops to grind downhill and join the Balchen.

"How far do you think?" I asked Cain, nodding toward Bird Bluff.

"Dunno," he said. "Maybe a couple hundred meters, maybe a kilometer or more."

Didn't matter how far—plenty of fuel for what looked like a short drive. I could do a mile or so on a machine, but not forty. "Let's head for the west side of the scoop, see if we can climb down into it from there— should be less steep, softer snow," I said.

"Okay, fine." We started off.

Cain led. I followed on my machine about thirty yards behind to his left. We approached the front of the wind scoop a hundred yards or so back from the edge. *Why go this way?* I wondered. *Why not head to the west side like I said? Did he ignore me again?* The snow cover we drove over proved thinner here; blue ice of the Balchen Glacier poked through in places. I felt my machine slip on ice patches.

I remembered an air photo taken here. The snow surface had a gentle slope toward the scoop, then rolled over, a sudden drop-off that steepened to the bottom—couldn't see the floor of the scoop from our location. I looked off to the side of my machine at patchy snow a few inches thick at most. Cain and I got closer to the scoop; we drove over more ice than snow. I maneuvered right and left to stay on snow to keep traction, not skid on the ice.

Ahead of me, I watched Cain drive off a snow patch onto the blue glare ice not far from the front of the scoop. *What the fuck?* That was odd; we were on a slope here. We could lose our grip. I stopped on a last small patch of snow, tense, frustrated that he took this route. I waited to see what would happen to Cain. He drove on. No sense in calling out— he'd never hear me and he couldn't see me.

Cain's snowmobile fishtailed; it lost traction. He turned to the right, slid sideways down toward the slope. I watched mesmerized, helpless. He got his machine pointed upslope but lost forward motion. He started to slide backwards down the slope. Cain and his machine slid toward the front of the deep scoop. He gunned the snowmobile engine. The tread belt spun. Studs on it meant to grab the ice instead carved into it, threw up a rooster tail of ice chips behind. He and his machine contin- ued to slide backwards down the slope.

"Hey, jump—jump off," I yelled. "Jump now, bail out." He couldn't hear me. I stared. Cain jumped off, landed on the ice hard; he looked stunned, frozen. The snowmobile gained speed and slid backwards—over the edge of the slope and out of sight. Cain lay on his back; he continued to slide feet first.

"Arrest, goddammit, arrest," I hissed. He turned onto his left side and slid. "Arrest, arrest, come on." Cain rolled on his stomach, popped up to his hands and knees in the arrest position, slowed and stopped. My heart pumped hard; now what?

Cain turned over and sat up on the ice, looked at me about fifty yards away.

"Are you okay?" I shouted.

"Yeah. Can you give me a hand, come over? Put on your crampons," he said. All his gear had gone down the scoop with the snowmobile. I clipped my crampons onto my boots and crunched over to him, bit into the ice with each step, helped him up. We made our way arm-in-arm over to a snow patch—he couldn't cross over an ice slope in his plastic mountaineering boots.

Cain and I stood together on the snow patch—no discussion of what just took place. He seemed okay. I felt amazed that Cain made it unharmed.

"Let's see if we can get down there," I said.

Near the cliff, I looked down at Cain's machine at the bottom of the scoop a hundred feet below, on its side in a pile of rocks, a fallen horse.

That could have been Cain—injured or dead.

"We can get down into the scoop here," he said. We climbed down, found good footholds in the deep snow drifted against the rock.

I heard a humming sound. We picked our way across a field of boulders at the bottom of the scoop, over to Cain's machine.

"Hey, this thing's running," I said. The bashed-up snowmobile lay there with its engine at idle—*putter, putter, putter.* We both laughed at the irony. I thought of Woody Allen in the movie *Sleeper* starting up a two-hundred-year-old Volkswagen.

Cain and I inspected the damage. The machine had a broken windshield and front cowling, a bent handlebar. The two of us got to one side and pushed the snowmobile up onto its treads.

"Let's clear a path outta here, drive up that side we came down," Cain said, pointing to the west side.

We picked up rocks that lay on the snow, cleared a narrow path to drive through. Cain climbed onto the snowmobile and threaded his way out of the rock pile to the cliff face at the west edge of the scoop where drifted snow had accumulated. He drove up the snowy slope to the top; I climbed up the snow drifts and out of the scoop.

At the top, I got on my machine, yanked the cord, started it up. I drove back with Cain the mile or so to camp. *What's Cain thinking about his accident?* I concentrated on my relief—no injuries. Didn't feel like chewing him out. *Glad he's okay, his machine still working, but I lost a workday. Not a lot of those left.*

Back at camp, we inspected his snowmobile then ducked into our tents to wait for Tucker and the rest to return.

I heard them drive up sometime later, stuck my head out of my tent, saw Cain and Tucker talk and point as they walked around the damaged machine. I joined them. The three of us stood by the snowmobile, looked at the damage. What did Tucker think? What did Cain think? Tucker explained how he would do repairs. He had tools and a few small pieces of lumber he had brought. I pulled him aside. Cain left.

"So, you can fix this. Great!"

"Yeah. Lucky it's not worse off," he said. "What exactly happened?"

"He drove onto glare ice and lost control. What do you think?"

"About what?"

"I'm surprised he did that."

"Probably thought the snowmobile cleats would hold. It's an accident, Bruce."

"Now we know."

"Yeah, and so does he."

I never heard the rest of us talk about this close call. We could have lost Cain and the snowmobile. We counted on Cain, our mountaineer, to be perfect. I realized he was a human like the rest of us. We all faced uncertainty. Hard to say what was a mistake and what was a chance encounter with unseen menace. I had to remember that at this point in our expedition, all of us were tired. All of us were prone to short tempers and mistakes. I hoped we'd stay safe. I hoped we'd leave here unscathed.

CHAPTER 25

Black Flags

This turbulent silence, the sprawling ice, and the occasional sharp gusts of wind warn that eventually you will make a mistake. The threat is babbled endlessly, as if Antarctica were a lunatic.

—**NICHOLAS JOHNSON**, *Big Dead Place*

Reece Pass Glacier, January 6, 1990

I stood in the quiet on the hard snow, tilted my head back, and examined the brilliant cobalt Antarctic sky. I thought back to the violent, blinding storm of last week and smiled with relief. Fine and warm today meant no need for our beefy, down-filled red parkas, the Big Reds. Windbreakers did the job. I reflected on how nobody is born ready for Antarctica. Maybe five humans had ever come to these mountains, remote even for Antarctica since discovery fifty years before. I learned the reasons for that: because it's hard to get here and harder to work here, and it's dangerous too. Antarctica is a place where mistakes are not forgiven. The six of us were here anyway.

I joined Cain as we loaded our snowmobiles with gear for a second attempt to sample rocks at the cliffs of Bird Bluff. He almost bought it there yesterday. I looked up the nearby Reece Pass glacier that cut through the Fosdick Mountains. The glacier flowed down toward our tent camp. The Reece undulated like a fat white snake. That meant the surface was cracked, that the glacier held deep, deadly crevasses, hidden beneath bridges of wind-driven snow no more than a foot or two thick.

I watched four of my team making their way up the glacier on their way to the south side of the Fosdick Mountains. Above Cain and me, about a half-mile away, I could almost make out each member of the group. They'd stopped their snowmobiles and now stood on the snow near a few black flags which marked concealed crevasses they discovered two days ago.

High-stakes on-the-job training had crossed my mind as I watched them leave camp. Tucker led the party of geologists, driving a snowmobile that towed a sledge. Chris rode the sledge that Tucker towed along with Steve. She had set up their trip today. Dave drove the last snowmobile in the train.

Chris' scream struck Cain and me from across the Antarctic snow.

"Tucker! Tucker! Oh my God! Oh my God! He's fallen into a hole!" Cain and I exchanged looks. I turned toward her sound. My body rigid, I held my breath. "Tucker! Tucker! Get out! He's fallen into a hole!"

My pulse raced. "What's going on?" I knew, but I hoped I was wrong.

"Let's look," Cain said.

Chris screamed again for Tucker to get out. Get out of what? Cain and I pulled out our binoculars, scanned up the glacier, hunted for details from what had been four members of our team. I could see only two figures. I tried not to be frightened.

I continued to stare through my binoculars, then identified Chris, and whom, Steve or Dave? I wished we had walkie-talkies, but we didn't.

"Two of them down a slot," Cain said, his voice flat while he examined the distant scene. I felt my ears burn. Tucker was down and someone else too. How? Weren't they roped-in? Had my worst nightmare happened? Were two of us dead in a crevasse? Panic threatened to take over. *We are alone. Help is eight hundred miles away.*

"Oh no, oh no—who's down? Who's down besides Tucker?" I turned to Cain—he lowered his binoculars, looked at me. "We've gotta get up there," I shouted. I pulled the starter cord on my machine. Cain grabbed my left shoulder.

"Steady, Bruce, steady—stay calm." Cain's face showed no expression. "I'm going."

"What?"

"You stay here. Stand by the radio to call for help—wait until I say so." Cain started his snowmobile and sped away over the snow.

I watched Cain race up the glacier to the spot where the accident happened. I stood helpless. On his order, "stay here," I went no farther. I knew the emergency rule: don't act if you could increase the danger. That glacier was laced with hidden slots—maybe one for me.

I stayed in place and tried to make out the actions up the glacier, but I couldn't hang on any longer and do nothing. I started my snowmobile and drove to the base of the Reece, then stopped. Now close enough to identify people through binoculars, I recognized four figures on the slope above me: Cain, Steve, Chris, Tucker. Tucker had escaped from the crevasse; he must have been roped. Steve and Chris stood still next to a snowmobile; Tucker and Cain crouched on the bright snow, looking down. One man missing—Dave. *Is he hurt, unconscious? Not on a rope?* I watched, listened, hoped.

I waited, with time frozen like the ice I stood upon. Frustrated, frightened, I shouted up to them.

"Should I radio Mac Center, send a Mayday?" *Four hours for help to come. If Dave's alive, he'll soon freeze.*

Tucker yelled back to me, "No, negative." I knew the consequences of making that call. An entire Search and Rescue effort would mobilize people, aircraft—all of Antarctica would stop. Other people would take risks to get here. We had to be sure we needed aid.

I didn't want to stand there, a powerless, distant witness to a tragedy.

"I have a towrope; I can bring it up," I shouted.

"Bruce. Stay there," Tucker yelled. I heard the tension and message in his voice. *Don't amplify this crisis, Bruce.*

Through my binoculars, I watched, my breaths shallow and rapid.

Tucker shouted as he looked down at the snow, "Do you have your jumars?"

He was shouting at Dave. Dave was alive. Tucker was asking him if he had the clips to allow him to climb up a rope. Was Dave injured? Could he get out? Would Tucker, maybe Cain too, need to go down into the crevasse and get him? Was he wedged in a slot, trapped? Would he die, his body frozen and never recovered? Did Dave have on his Big Red when they left? I tried to remember. He always wore it—maybe so. That would give him more time to live.

Early in the expedition, after the first blizzard we endured, and after one of our simple one-pot dinners, Dave and I relaxed in our tent and chatted about adventures and dangers we might encounter. We faced each other across the tiny one-burner stove and wooden boxes that lay between us along the center of the tent floor. He struggled to get his large body into his sleeping bag, propped himself up on his left elbow to face me, pulled on a dark blue knit cap over his partly bald head. We

Tucker shouts to Kimbrough in crevasse. (Photo: Steve Richard)

talked about trials of early explorers like Scott, Amundsen, Shackleton. We agreed we wouldn't have it that bad, but we found ourselves on our own in a place full of essential peril. Dave summed up his thoughts.

"I won't die here, Bruce. I know that."

I waited at the foot of the glacier that held Dave captive in its blue jaws. The infinite sky, the silence, the perfection of the snow—all lost on me. I stood, captured by alarm within this no-longer-magical place. I waited hours, watched through my binoculars. I listened. I paced. I hoped. *Dave can't die. This can't happen to us, alone out here. No help will come for us. Everyone warned us. Did we listen? My fault—this expedition; my idea.*

After a couple of hours, through my binoculars, I could see two figures on the glacier crouch down on the snow surface. Cain and Tucker struggled. They were helping Dave. Dave made it back to the surface. They got him out.

Every breath that I ever took in my life escaped me.

I choked a sob, watched them organize themselves. Dave sat on the sledge—I couldn't tell if he was hurt. I couldn't make that out. I had to wait in place. Wait.

I watched Cain and Tucker probe the snow for more hidden slots, to find a safe exit route for Dave, the sledge, and them. Chris jumped on a snowmobile, sped down the glacier toward me. She pulled up at my side.

"Dave's going to be okay." She rushed to embrace me.

"Thank God."

"Yes."

We held each other and sobbed for a few seconds.

I followed Chris back to camp, relieved, but alert with adrenaline. Tucker returned, towing the sledge with Dave on it. They came to a halt by the tents. I strode over and put my arms around Dave. He sat on the bundle of sledge cargo. "Good to see you, Dave."

Tucker joined our group hug. "Doesn't get much closer," he said. His truth tipped the scales. Dave and I hit our limit, let go and cried—tension, stress, raw fear gushed out. Cain stood in the background, looked on with his poker face, coiled his climbing rope in his left hand.

Dave Kimbrough safe on snowmobile after rescue. (Photo: Steve Tucker)

* * *

Our orange tent shone bright and cool, the air still, peaceful. I made tea. The tent interior now scented with brewing Earl Grey, Dave lay on his bag, an injured knee propped up. His mood joyful, he celebrated life. I needed to hear his story, why two of us almost got killed, here, alone, at the end of the world.

"Okay, tell me, Dave. Tell me what happened." I faced him and struggled between my relief and happiness.

"Well, I almost died."

"Yeah, Jesus."

"We wanted to cross over to Richardson again—retrace a route up the Reece glacier we made a couple of days ago with Cain. Remember? Cain found a covered crevasse on the glacier that day. He marked it with crossed black flags. We followed our old tracks—to the flags.

208

"A few yards short of the flags, we stopped. Tucker wanted to probe and test the strength of the crevasse bridge."

"The bridge was across your trail?"

"Yeah. He clipped his rope to his snowmobile and walked off. Chris, Steve, and I watched him. We stood next to his snowmobile."

"Were all of you roped-in?"

"We tied-in to the sledge towrope to travel up there, but unclipped to stand next to Tucker's machine, you know, to watch him," he said. "We checked out that spot Thursday—seemed okay."

I stayed silent, stone-faced, but knew that on this fine day, we let down our guard. Not roped standing on that glacier. They fucked up.

"Tucker went over to the flags. The snow bridge failed, had to be weaker with the warmer temperatures today. He fell in up to his chest; his rope held him. Tucker struggled out and came back to us.

"I thought it'd be cool to get a photo of Tucker in the slot, on his rope, head and shoulders sticking out, so I asked him to get back in.

"Tucker went back to the bridge, fell through deeper. His rope stopped him below the surface—outta sight.

"When Tucker went down, I felt the snow give way under me. That's when I fell.

"I thought, this is it—I'm gonna die. I banged off the sidewall of the slot, slid down a slope, and stopped on a snow ledge."

"Holy shit, that's nuts! Not roped—sloped sides, not a vertical slot. Unbelievable. You'd still be wedged in there, frozen." *He would have died in there and we couldn't have gotten his body out.* Who would've told his family, his mother, his fiancé? Me. I thought of that nightmare for a second. I took a gulp of tea from my mug. *This work is way too danger-ous. Cain dodged a bullet yesterday in that snow scoop at Bird Bluff, and now this. We're not working on a cure for cancer. What am I doing?*

"Yeah. I couldn't believe my luck when the ledge stopped me. I fig-ured, hey, I'm still alive, so I got out my camera to take a picture, but it got smashed when I bounced off the wall of the slot."

We both laughed. "You're always a scientist—take evidence," I said.

"I tried to figure out my situation. Not much light, dim and blue in there, and quiet, cold. I wore my Big Red though. The others didn't. That bought me time. I heard my heartbeat and my breath. I could almost stand, but my knee got whacked when I hit the wall."

"Fuck, man."

"I couldn't see the hole I fell through. I didn't know how deep I was."

"How much rope did it take to get you out?"

"A full pitch."

"Shit, that's maybe a hundred feet."

"Yeah. Crazy, huh?"

"What did you think would happen next? I'd crap my pants," I said.

"I figured the guys on the surface would get busy and try to get me out," Dave told me. "I could hear Tucker yelling. I yelled back, but I could tell that wasn't gonna work."

"That's weird. You could hear him, but he couldn't hear you?"

"Might've been asking me if I could get out on my own. Wasn't sure. Then I heard Tucker's police whistle, so I grabbed the whistle from around my neck and blew hard as I could," Dave said. "I heard Tucker whistle back—that felt great. So, we figured out yes and no signals, and Tucker knew I could climb out."

"You're so fucking lucky."

"I know. So I waited for Tucker to come down, or a rope or something. Then the rope slid down to me. Tucker had weighed it down with ice screws so it could reach me."

"That was smart of him."

"Yeah. When I saw that rope, I knew I would live." Dave opened his eyes wide when he told me that, nodded his head, then smiled. "I tied-in as fast as I could."

What could I say? I felt my heart rate increase, my skin tingle.

"I crawled my way up the sloping part of the slot, kept the rope tight," he said. "My knee hurt, but I had to crawl."

"Your knee, how did you get it to work?"

"My knee got banged up bad, it hurt like hell, but no choice, man."

Dave crawled over the pain of his injured knee.

"I made it upslope to where the crevasse looked vertical. I could see the hole I fell through with Tucker's head and the blue sky above. I got excited when I saw him, but he looked far away, forty or fifty feet.

"I gripped that rope tight. I knew I could do it, no choice. I clipped my jumars on the rope, started to climb up, real slow. It hurt each time I stood on my bad knee."

"How long did it take you to climb up?"

"No idea. Maybe an hour, two? Don't know. But I shimmied up the rope 'til I could touch the surface, my head just out of the hole."

"I can't imagine what you thought or how you felt."

"Man, grateful that we trained to use jumars for starters. Then Tucker and Cain grabbed my parka, dragged me out of the hole onto the snow."

"Jesus, Dave. Those guys hauled you out? You're not a featherweight."

"Yeah. They were pumped. I lay on the snow for a second or so on my back, took in the rush of being alive." Now he broke into a big grin and a low laugh.

Almost two hours after Dave tied himself to the rope, he came back into our world of sky.

"Fuck, Dave. I was freaked the whole time, not knowing, thinking you were killed. Fuck, fuck, fuck!"

"I thought I was finished, Bruce." He looked away, raised his eyebrows, and shook his head.

"Dave, we need to figure out how to not let this happen again."

"Yeah, we still have a lot of work to do."

CHAPTER 26

Up the Ochs Glacier

Fear, to a great extent, is born of a story we tell ourselves...

—**CHERYL STRAYED**, *Wild*

I didn't belong here—nobody did. Now I knew what that meant—we could die in this ominous land. Before today, the notion had been a bad fantasy that surfaced on occasion. Now I knew for sure.

"Let's meet and discuss what happened today, what we'll do next," I said to the team gathered around the sledges after Dave's rescue.

Chris, Steve, Tucker, and Cain crawled through our tent opening, found seats on Dave's bag and mine. Dave squirmed to get comfortable, his knee injured. Faces were blank. No one spoke. What to say? What did they feel? Stunned? Exhausted? Freaked out like me?

I began. "We got lucky. Dave's going to be okay; we didn't need to call for SAR or a medevac." I scanned their blank, silent faces. "We must keep this accident to ourselves. No one can know; it doesn't leave this tent."

"How come?" Dave said, his face questioning.

"Because NSF might decide we're not up for this, not allow us to come back next year." No reaction.

"The weather's getting too warm for safe glacier travel—I think. We can't count on crevasse bridges to hold up. I'll bet we crossed a lot of them, didn't know that, and got away with it." People looked at their hands, their boots. *Why am I not getting any feedback?* "We need to get back to the Depot and work the south side of the Fosdicks—no crevasses there," I said. "That's where you were going anyway, crossing the glacier to get to Richardson on the back side of the range." Some nods, no words, passive faces. "We're pretty much done on the Balchen too." No objections. "I'm not getting any input; no comments?" I paused, then continued.

"We'll move camp later tonight, midnight, when it's coldest and bridges are stiffest, back to the Depot." The group shifted in their seats at this idea, like they sat on thumbtacks.

Tucker spoke up. "We need to rest. Dave's hurt. It will be cold enough in the morning; bridges should be stiff then." Murmurs of agreement. Watching them, I could sense exhaustion. Only Cain was hard to read.

"Yeah, thanks, Tucker. Okay, probably better to move after rest, but before it warms up."

January 7, 1990

The six of us left our geology campsite at Bird Bluff late morning. Our train of snowmobiles and sledges headed down the wide, gentle slope of the Balchen Glacier, four snowmobiles and six sledges. The destination: our cache at Depot Camp. To get there, we had to retrace our trail up the Ochs Glacier that we descended a week before. Today's route followed the south side of the Balchen, close to the gray cliffs of the Fosdick Mountains. Snow had drifted deep in places and the ride was smoother than usual, but slow just the same. Dave and I each drove one machine that pulled a heavy, loaded sledge. Cain took the lead snowmobile towing two sledges. Steve rode brakeman on the second sledge, standing on the rear, gripping vertical handlebar posts. Tucker towed two sledges with Christine riding brake on the last one.

We left camp late morning, later than planned, the team still wiped from their ordeal. The weather had looked fine, but soon the signs turned bad. Clouds came in; the wind shifted to come from the feared northeast and increased, indifferent and with familiar menace. I looked at my watch. Its barometer showed a large pressure drop. I felt frozen air inside my parka. We would be in for it soon. I thought back to the storm at Birchall. Cain stopped the train near the cliffs. The rest of us drove up to him.

"Let's gather the sledges and snowmobiles to make a wind break and have lunch," Cain said. We did that. Sitting on the snow behind the sledges, hunched up to get out of the now icy, bitter wind, we lunched on hard salami, frozen cheese, and hot tea. The sky showed a turbulent gray; snow danced around us in gusts. We tried to get low and small to preserve warmth. The wind increased. Blowing snow pelted my face.

Cain approached me. "Bruce, this weather looks grim. We could be in for an epic," he said, loud over the roars of wind. "We should stop here and wait this out; could be hurricane-force at the top of the glacier." This gave me pause. Cain never exaggerated.

"We're only a couple hours from Depot. I don't want us pinned down here for days; we can make it."

"Not a good idea to move in this. What do the rest of ya think?" Cain turned to face them.

"Yeah, let's go," Steve said. No hesitation. He wanted to work, not wait. He stood now and faced the wind that whipped through his scraggly black beard.

I looked to Tucker, who didn't give me a clue. It was hard to read Chris and Dave; they sat on the snow, covered up with their parkas and goggles, trying to shrink into a ball behind a sledge.

I wanted to get back to replenish our supplies. We could be stuck here for days. We needed stove fuel. No fuel, then no water to drink from melting snow. I had packed extra fuel when we left the Depot, but Cain unloaded it, maybe to save weight and space, maybe to show me I'm a pussy.

Besides, Dave's crevasse plunge yesterday had traumatized the team. He almost died. He was injured. We needed to recuperate. We needed to heal in a place we imagined as safe—the Depot. Never mind that it wasn't true.

"Looks like we're going," I said.

Lunch finished, we got underway. In an hour, we arrived at the base of the Ochs Glacier. Cain stopped ahead; Tucker, Dave, and I pulled up to him.

"What's this?" I looked at an icy swamp and stream of water about sixty feet wide that lay across our path—formed since last week. I could hear water running, even with the wind.

"The glacier's melting. We're at a lot lower elevation here," Tucker said. He looked surprised, like me. I remembered that Pete warned us about this in Christchurch. Watch out for January melting low down and so close to the coast.

"We got to cross this; no way up otherwise," Cain said.

"What? How deep's the water?" I imagined we could lose a person or a sledge here. I looked around to see who else reacted. Didn't matter— we had to cross.

"Goin' to check this; can't be deep," Cain said. He pulled out a couple of ski poles and stepped into the stream. *Hey, what the hell's he doing?*

"Stop. Don't go." Cain ignored me. He picked his way with care, probed, tested for thin, false bottoms, made his way across. Did his boots and feet get soaked?

"We can get across," Cain said. "Run for it fast as you can go."

"All right—I'm towing just one sledge, so I'll go first." *Why'd I say that?* I needed to keep my fear under control. *Don't give a shit*, I told myself. I gunned my machine, headed into the stream, and splashed across, sledge in tow. Success—I stayed dry. Cain, Dave, and Tucker followed. Tucker's two sledges got stuck midstream, and one toppled over on its side into the flood. Chris, Tucker, and Steve waded into the swamp to right it and push the sledge out. We needed to get warm real soon— wet feet freeze. *Jesus, wet feet in Antarctica; what's next, wet bodies?*

Across the stream, we gathered and looked up the Ochs. About a mile wide and a few long, the glacier flowed many hundreds of feet downhill through a canyon of vertical rock walls, gray and black, that thrusted some two thousand feet above the ice surface. Snow petrels circled high up on the cliffs, buffeted by gusts; like ghosts they glided. The birds flew as though this blow presented nothing special—a free ride on powerful updrafts.

"Look at the blue ice." Tucker pointed to the higher, steeper blue parts of the glacier. The wind had blown away more of the snow cover. Less now than when we went down this route. We wouldn't be pulling these sledges up slopes of glare ice. "We'll need to pick our way uphill, move between snow patches," Tucker shouted.

If I started to slide down backwards, I'd bail out. Now I really didn't give a shit.

We moved up. Cain led with two sledges and Steve in tow. I followed them. Our path zigzagged up from snow patch to snow patch. I saw the six-foot bamboo poles with trail flags we'd planted a week before—blown down. We climbed.

Chris crosses meltwater streams on Ochs glacier. (Photo: Steve Richard)

The higher we reached, the stronger the wind. Our train broke out of the canyon that held the glacier, then onto the Chester snowfield. The wind smashed into us like a colossal, furious wave. Blowing snow obscured my vision. Snow froze to my goggles; it was hard to see Cain and Steve ahead. *I'm dead if I get separated.*

The trail flags gone, our tracks in the snow from a week ago blown away, I glanced behind me. Tucker, Dave, and Chris had stopped. The wind had toppled a sledge that weighed hundreds of pounds. Steve looked back and signaled Cain to stop while we regrouped.

Like Cain predicted, we found ourselves in an epic. We could either circle the wagons here or search for the Depot. Didn't matter; we'd never practiced tent setup in a gale. Might need to dig snow caves. Snow was so hard we needed pickaxes we didn't bring.

Then Cain started to drive like he knew where to go. Through horizontal white streaks of jetting snow, I could just make out Steve's shadow ahead of me. He stood on the rear of the sledge. Cain drove where Steve pointed.

No choice, I followed. Dave and Tucker followed me. After twenty minutes of back and forth, Steve found the way in the snow gale, and we came upon the Depot and our cache of supplies. Unbelievable. A warm relief calmed me despite the assault of the storm. How did Steve see our way? I was almost blind. Our team crowded around, shouted at each other over the wind—we made it. I slapped Steve on the back. "Thanks for navigating, eagle eyes," I said.

Now to get tents up—no easy task with wind and snow blasting us at over forty knots. I felt snow inside my parka.

Cain grouped us together. "We can do this. Get four people on each tent, two with lines on the peak upwind and two at the skirt downwind. Let the wind pull up the tent peak."

We had tied the Scott tents to the top of the sledge loads, always packed up last after we broke camp so that we could set them up first. Dave and I sprinted to a sledge to untie our tent. *This had to work.* Survival School had told us how to raise a tent in howling winds, but we'd never done it under those conditions. If other people did it, so could we. We had to. I was too cold to feel anxious.

"Keep the tent peak upwind or it will fill up and blow away outta sight!" Tucker yelled. We moved the heavy pyramid tent into position and laid it on the snow. Its peak pointed to face the gale. Chris and Steve came over to help us.

"Hey, Dave. What's happening?" I looked at Dave who stood next to me.

"The wind's dyin'. I can't believe it."

"You're right. Quick, this is our chance."

Chris and Steve rushed to grab their tent. We unrolled our tent, dug the postholes, spread out the tent, and set it upright in the holes. The others started to set up their tents. We grabbed pipe stakes and our rock hammers, pounded in the pipes, pinned down the tent skirt on all sides. We moved fast.

"Dave, the guy ropes—you take the back side," I said. We attached the guys and in unison pulled tight the front and back ropes. "Now the sides," I said, and we did the same.

"Bruce, shovel," Dave said, and tossed me one. We shoveled at a breakneck pace, piled snow on the skirt to bury it under several feet of snow—to make it bombproof. If Dave's knee hurt, I couldn't tell. He shoveled. The wind stayed down. We unloaded sleeping gear and food boxes from our sledge. I shoved the floor tarp into the tent.

"Get inside, Dave," I said. I threw the wooden boxes and duffels into the tent. I dove into our tent and collapsed, exhausted, on the yellow tarp that we'd stretched over the snow. We both lay there quietly. My heart drummed in my chest—my breath came in gasps. I stared up at the orange peak of our tent eight feet above my head. I controlled my breath to calm down; in, out; deep, slow; calm, ease, smile, release…. *We're okay now. Those wet feet can dry out.*

Next moment, the canvas tent walls shuddered and snapped like gunshots. They vibrated and began to hum a low moan. The thick wooden tent poles flexed.

"Bruce. The wind. It's back."

"We're good now, Dave; we're solid." I said what I had to say, true or not. In this land of hypnotic chaos, luck mattered. I hoped we still had some left.

CHAPTER 27

Pull-Out

No man should go through life without once experiencing healthy,
even bored solitude in the wilderness, finding himself depending
solely on himself and thereby learning his true and hidden strength.

—**JACK KEROUAC**, *Lonesome Traveler*

Depot Camp, January 7–15, 1990
Tuesday, January 9

I lay on my back on my sleeping bag, just a few inches above the snow surface; my back felt comfortable and supported. I stared at the dull orange tent peak, the canvas pyramid eight feet above my head. Its walls were peaceful and moved with only soft ripples; the gale we traveled through had died a few days back. Wet gloves and socks hung by clothespins on a line strung in a square around the interior of the peak. I reflected on a crucial lesson I'd learned from repeated teachings: luck is important in Antarctica. We had arrived at the Depot two days ago, Sunday, January 7. I thought back on our ascent up the Ochs Glacier in the snow gale. We found our cache in that blizzard by luck as well as skill. I thought about Dave's crevasse accident, Cain's snow scoop fall. *Why are we getting away with this? Why aren't any of us badly injured or dead?*

Pull-Out

Monday, we had started to pack for the two Pull-Out flights. Dave and I had moved the scheduled Pull-Out up a week in anticipation of delays. Chris and Steve had been annoyed by our decision. They wanted to work, stay longer. I got that and appreciated their motivation, but Dave and I needed to get back to our jobs in the world. At this point, an extra week or more would be a big problem. Dave, Tucker, and I would go out first, January 10. Chris, Steve, and Cain would stay and work five more days to January 15. We boxed our gear and supplies, strapped them to aluminum US Air Force cargo pallets four feet long on each side, covered the loads with cargo nets, and lashed them down tight. These would be winched up the tail ramp of the Herc at Pull-Out. We had not packed air cargo before, but it looked to me like we did a great job.

Dave prepared his watermelon-sized rock samples, smashed them into smaller pieces with his sixteen-pound sledgehammer, and packed them in rock boxes. He made a steady sound of *crack, whump* as his hammer broke the rocks, then struck the hard snow. Dave didn't complain about his knee injury, but I thought I saw him limp.

Thinking about our accomplishments helped me feel better. I had drilled about one hundred cores to work on in my lab. I'd determine if any twisting or shuffling of these ranges had occurred during their formation millions of years ago. Dave had sampled every outcrop we visited. His lab work would get us crucial age control. Steve and Chris, from recognizing key metamorphic minerals in the Fosdick migmatites, had a rough outline of a history of that range. They also found evidence of multiple episodes of tectonic deformation, later ones overprinting earlier ones. Dave's age-dating promised to pin down the timing. Eventually, we'd put together the whats, whens, and whys for the breakup of Gondwana in MBL and could compare this to what was known about New Zealand and surrounding plateaus—but not before we explored north to the Phillips Mountains to fill out the story. That was our goal for the following season—more of the same in another range.

Our schedule was abandoned in the face of abrupt reality—Antarctica was not consulted. So, as I expected, we waited, and my tension mounted. What we had been warned about weeks ago now appeared to be in progress: be prepared to wait for days or even a couple of weeks for Pull-Out—the word from the OAEs. Now the weather sucked—overcast, warm, intermittent fog. I became anxious.

* * *

Dave announced his idea. "Hey, Bruce, we're havin' a competition, the Fosdick Cup, a snowmobile rally," he said with a big grin. Dave sensed the tension of standby-to-standby waiting and organized an event to build morale. The snowmobile rally would be a contest for the fastest time in a slalom course and a drag race. He told us he set up a costume contest too. Who could look the most ridiculous wearing the clothes we had brought?

He planned these contests on his own. This reminded me of the Dave I knew in Santa Barbara: playful, a bit off-center. First time I had seen him show this on our expedition. I felt surprised by his upbeat mood considering he almost died a few days ago. Maybe that was the explanation—he didn't die, so he was celebrating.

Dave named off the prizes: "First is a bag of Hershey's Kisses; second, one can of Polar Crab Meat; third, a box of chicken bouillon. Whoever comes in last gets a bag of Mountain House freeze-dried chicken stew, and as a bonus, a can of Dinty Moore stew," he said, laughed. We hated that stuff. "I came up with the last-place prize with you in mind," he said. "You're the old guy so nobody's got money on you, ha, ha, ha."

I needed this diversion. I crawled out of the tent. Tucker, Cain, Chris, and Steve had dressed for the costume competition before I could think about it. Tucker and Cain wore their long underwear and briefs and motorcycle helmets, Chris wore a sheepskin toga she had made from her sleep kit over blue long johns, and Steve strung himself with Christmas ornaments we had brought. They did look ridiculous. I laughed, took a photo. But I found myself caught flat-footed—no entry. The contest result: a tie. I got a pass.

Dave had laid out the slalom course with bamboo poles with red trail flags. Six switchbacks covered an area on the snowfield around the camp. Each leg was about thirty yards with one-hundred-eighty-degree turns. The race path traced a round trip out through the course and then back to the start. "You go last, Bruce," he told me, laughed more. His attitude made me pay attention to the event, to find a way to win the race.

I watched the others, one at a time, gun their snowmobiles through the course. They overshot each course pole, had to back up, go forward, back up, turn, race off. I made my plan. My shot was now. I gunned my snowmobile, approached the first pole, then crossed my outside leg to the inside running board that faced the turn flag, stood on the inside, tilted the machine like a ski to edge around the flagpole. I switched sides each turn, cruised through with no dead-ends, no stopping and backing up—beat everybody by half their time. Dave shook his head and the others laughed.

Dave's drag racecourse put us one-against-one, side-by-side. I raced last. I watched the others gun their engines at the start, lumber off, and race to the finish about a hundred yards away, where Dave stood with a flag and stopwatch. My turn. I lined up against Tucker, gunned my engine full power but held the brake on, the clutch smoked. Dave dropped his flag to signal the start. I released my brakes, launched off, and left Tucker behind—beat him easy.

"Who won the drag, had the best time of all?" I asked Dave. I knew the answer. He looked at his stopwatch.

"You did best, dammit." He shook his head again. Dave's admission gave me a boost. I got to laugh last and eat the bag of chocolate Kisses.

January 10-12

Our Pull-Out flight, event A421, was scheduled for Wednesday, January 10. Dave and I reported the weather every three hours on Tuesday, then every hour—the torturous hourlies. We took shifts as the flight launch time approached—fourteen-hundred hours in McMurdo. The flight would first go to the WAVE camp an hour east of us, airdrop them supplies, then come back west to get us—should be here early evening about nineteen-hundred hours. The day started perfect: clear, calm. Optimism showed in cheerful banter and hasty footsteps as we prepared.

"Hey, Dave, looks to me that our plan will work—split the party with one flight today and another next week," I said. Then I watched the temperature climb. Yesterday, Tuesday, by mid-morning, the temperature read +9°Celsius (+48°Fahrenheit), a heat wave. Melted snow meant a wet mess of slush and a sticky surface for aircraft takeoff—that could be a bad day.

"It's gettin' warm, Dave. That means fog."

"Yeah, probably; that's tough, man," Dave said.

By that Wednesday afternoon, fog came up from the ocean onto the Crevasse Valley Glacier and engulfed us. We reported the weather— visibility nil. I asked Mac Center if we could be put on standby. No go—snowing in McMurdo, flight cancelled. They must have known snow was expected in McMurdo. Why had they planned to come here?

We knew there were multiple moving parts to get a flight out to us: good weather in McMurdo, better weather at our location in the unfamiliar mountains, availability of aircraft, and a rested flight crew. Monitoring the radio, we learned that flights were made for other field party Pull-Outs, transported folks to Christchurch and fuel to South Pole Station. Patience, keep it together—KIT. I felt this sudden switch in the weather and flight cancellation marked a bad omen. The fog could stay here a long time—days, weeks. I worried. Every day tested our run of luck.

Thursday, we were told that our Pull-Out had been scheduled for the next day, Friday, event A429. Weather had improved, partly cloudy, so we had a shot at it. Chris, Steve, and Tucker had worked part of yesterday and then today at Mount Getz; no time to waste. Dave and I did hourlies for the flight. Friday arrived with low clouds, snow, fog—Pull-Out cancelled. Saturday, we were on for event A435. That flight also cancelled—low clouds, fog, snow. Three attempts failed over the course of the week.

"There were two good weather windows we missed, Dave. They scheduled us for the shitty parts."

"Not on purpose, Bruce."

"Now we don't have enough time to split the team."

Spirits faded. Dave looked quiet now, spent a lot of time in his bag reading. He could do what he wanted, I figured. *Why hasn't he lost it? How does he KIT?*

Sunday, January 14

I stared up at the peak of our tent, a promising orange glow—bright color, a good sign. The sun had come out at last today, Sunday. My Walkman, on its last batteries, played "On Your Shore" into my headphones. Chris had introduced me to Enya's music. Her song calmed me, reassured me. When the batteries started to fail, I could listen to speech but not music, as the tape speed dropped off while the batteries died and made the tunes sound basso. But I could still listen to my relaxation tapes.

I had become anxious; more than that, I felt stressed to the max, almost fearful. What did that mean? Then I remembered. Cain just missed sliding off a cliff. Dave fell into a deep crevasse. I—we—almost got lost in a gale. That was my decision. *Why did I put everyone in this situation? I can't handle another accident. We need to get out. I'll never come back here.* My journal entry for Sunday read, "I'm trying to put away fears of us being left out here, forgotten, lost in the NSF and USAP chaos—I know that's ridiculous."

Since Wednesday, five days ago, three Pull-Out flights had been scheduled and cancelled. I went over to visit with Tucker in his tent. I needed to talk. His tentmate, Cain, had left with Chris and Steve to do geology on Mount Richardson. I had an idea to run by him. I needed to KIT. I crawled into his tent; he had aromatic Earl Grey tea ready. I squatted on Cain's bag, faced Tucker.

"I'm thinkin' we're in a tight spot, not being looked after," I said.

"What do you mean?"

"They need to schedule more flights for a chance to catch these short windows of good weather. They don't get it, what it's like out here. In McMurdo, there's way more clear skies," I said.

"Why don't you ask for more flights?" he said. I ignored his sensible suggestion.

"Look how we were treated after our Put-In. The second flight promised hours later didn't show up for two weeks. They don't give a shit," I said.

"We have plenty of food and fuel," Tucker said to remind me, his voice even.

"Yeah, but I don't see an end to this fog. I'm stressed; we could be out here for weeks."

He raised his eyebrows, looked at me. "Bummer, dude," he said, tried to make the moment light.

"We could move the camp and a lot of our cargo inland, maybe a hundred miles, higher ground to get out of the fog, colder snow," I heard myself say.

Tucker sensed my loss of rational thinking. He turned to face me. "What?" He paid close attention now.

"Makes sense to me—get away from the coastal fog to a place where we can get a flight. Hell, we could cut trail to McMurdo if we need to," I said, surprising myself to spout nonsense.

"We're not desperate, Bruce," he said. "We won't be left here. Ask for multiple slots on the flight schedule, like you want. Remind them three flights have been scrubbed in the last week." He looked at me, almost, but not quite bemused, rather concerned that I might be too worked up. He saw I needed to calibrate, get grounded, to KIT.

"Yeah, you're right. I'll call in."

* * *

"Dave, did you hear that?" I clicked off the radio. He sat up in his sleeping bag.

"Hey! They're finally serious about getting us outta here. Three Pull-Out flights scheduled," he said, laughed like he'd won a poker hand.

"We've got to start hourlies this afternoon. I'll go first."

"I'll do night until about three in the morning. Then you take over; okay, Bruce?"

"Yeah, I'll be stoked for that."

Steve and Chris descending mountain. (Photo: Steve Tucker)

Monday, January 15

Early Monday morning, Dave woke me. I called in weather, left the radio on to listen to Mac Center and aircraft comms. "Event A four-four-two launched oh-five-thirty," I heard Mac Center announce. "Mission WAVE airdrop, Fosdicks Mountains Pull-Out Sierra Zero-Seven-Zero." *That's us.* I felt electric with contained excitement.

I looked at my watch; I did the math—four hours out to WAVE, an hour back to us, expect our Herc here mid-morning Monday. Perfect. I crawled outside, looked up at the sparse cloud cover. No fog. I could see about twenty miles across the Balchen Glacier north of us. I read the thermometer—plus three Celsius, way above freezing. Fog again? Wet snow—could be bad news.

We'd prepared our cargo in a line to be loaded. By early morning, everyone got out of their tents, stood, and waited—a lot of pacing, searching the sky, listening. I moved the radio out of my tent onto a rock box placed on the snow. Clouds came in. Blue sky shrunk to zero. *No, not more of this.*

A radio call came in. "Sierra Zero-Seven-Zero, X-ray Delta Zero-One."

"Go ahead, Zero-One," I said. Now my heart thudded.

The pilot came back. "We're near you, but the cloud cover's total. We can't see the ground. We won't drop through clouds; we'll look for holes," he said.

"Roger that." I didn't reveal my frustration. *All this waiting only to get skunked again?* The weight of our circumstance began to crush me.

"The clouds are above the peaks maybe a thousand feet," I said to Dave.

"Too low—not safe," he said.

"Yeah, not enough ceiling. You've seen the maps they use—no detail, don't show all the peaks." I gave Dave useless information, to have something to say.

For the next hour, we heard the Herc fly back and forth out of sight above the clouds, to look for a hole. The radio stayed quiet, no traffic; all of Antarctica listened. Dave, Tucker, Cain, Chris, Steve, and I waited. Dave and I stood by the radio. The rest checked, rechecked our cargo, wandered about. I felt group anxiety.

To the north, I noticed speckles of sunshine on a snow slope off the Phillips Mountains next to Block Bay, about fifteen miles from us—a hole in the clouds. *That might be our gift—wow.* I radioed, "Zero-One, we see a hole in the clouds, grid southwest from us, maybe fifteen or twenty miles."

"We'll go look." We heard them fly off. We waited.

Thirty minutes later, Dave said, "I see them. Look!" He pointed to the sky above Block Bay, gave a laugh of surprise. "I see their landing light." At first, I saw nothing. I squinted, then saw a tiny light.

"I can't believe it; they dropped through," I said. I looked around at the team. They faced me. I saw my emotions reflected in their faces: relief, excitement.

"Where are you, Seven-Zero; we don't see anything," the pilot radioed. *What? Now they can't find us?*

"I'll set off smoke," was my reply. I reached into my pack, pulled out a smoke bomb, and lit it off. Plumes of red-black smoke rose and drifted away.

"We see the smoke," the pilot said. "Coming your way. We'll do ski drags, then land."

To our west, clouds dropped lower, the mountains now obscured. Snow fell there, only a few miles away. *Snow. Stay away.* I agonized for the next twenty minutes. They circled the Chester snowfield, made two ski drags, banked to come in and land. The snow cloud moved closer.

The Herc touched down a couple miles away. Snowfall increased. I lit off two flares—the pilot radioed that he saw them, taxied the plane toward us. Cain pointed to the Herc, broke into a wide grin; I took his photo. They came to a stop between the mountains, snow cloud, and us, reduced the pitch on the props, and sat at idle. The tail ramp hissed open. Two navy crewmen came down, conferred with Tucker and Cain. I welcomed the harsh noise of the engines, the smell of exhaust fumes, pleasant now, the blast of the propeller wash. All demonstrated power that would carry us away to McMurdo.

Pull-Out Herc XD01 approaches Depot camp.

They were here, but the snow came in too. It fell on us. A few minutes later, they couldn't have landed.

We moved fast, like we planned. Took down our tents—that final step meant we'd leave. Our cargo pallets froze stuck to the snow. We struggled, got them loose, towed them to the Herc's ramp; its four engines were so loud I couldn't hear what anyone said. We gestured to each other.

Tucker yelled in my ear. "We need to sort our rocks, sledges, pallets to the rear of the Herc. The crew said they need to make the tail heavy to rotate up the Herc's nose, to get airborne off this wet snow." We rearranged our cargo. The crew winched our loads up the ramp into the hold, strapped them down.

Loads in the Herc, our team stood on the snow in our Big Reds. Rucksacks slung over our shoulders, we felt the sound of the engines, watched the tail ramp lift and close. The snow cloud came full on us.

We hurried to the left door of the Herc. Dave led; I approached last. I strode past a cargo net left on the snow for the second flight. That flight would pick up the last of our loads—when, who knew. I saw a manila envelope in the net, S-070 written on it. *Why is that there?* I grabbed the envelope, slid it inside my parka.

I stepped up the stairs to the aircraft door, turned to look back. Quiet snow fell—a whiteout soon. A good idea to remember this sight. For a couple of seconds, I accepted the privilege of my visit here. I knew I had to leave. I wanted to. I slipped inside. A crewman closed the door, then latched it.

Our team buckled into red web seats along the right side, inserted yellow earplugs. Ninety minutes to load; nobody could complain. Engines roared; the Herc broke away from the captive snow, lurched forward, and taxied for takeoff. No one spoke.

We lumbered across the snowfield to the southeast for several minutes, then stopped. The pilot applied full power; the Herc skipped along the snow, gained speed for takeoff. The plane shook, twisted, banged, hit the sastrugi. Ten seconds, twenty, thirty—the pilot cut back the power and stopped the plane. We didn't need to ask what happened. The Herc couldn't get enough speed on the sticky, wet snow to rotate up and lift off.

Are we fucked? The plane can't stay here. If the Herc can't take off, the crew will unload us and our cargo to get lighter.

After a minute, the Herc turned around, taxied again, I guessed to head back to the start—to unload? Five minutes later, we stopped at the remains of the Depot. Now we pointed in the opposite direction, facing the mountains. *How close?* The pilot idled the engines. I saw a crewman speak into his headset to the pilot. I looked back to the tail ramp. I waited for it to drop, open for our march down and exit.

The pilot applied full power. We lurched forward, skipped, and banged along on the rugged snow for about twenty seconds. I felt the nose tip up, down, up again. The Herc held this position, then left the snow behind—it reached into the air. Across the cargo hold, a young navy crewman sat, his parka hood off. I saw his face, tense. He puffed his cheeks, let out a big breath. I realized that my gut had seized. I let go.

In a few minutes, we cut up through the clouds. Sunlight entered the interior space through the few portholes. The pilot came back to meet his prizes. We had him take our photo. I framed a copy, a photograph I value—grins on raccoons. It made me smile at that remote moment. We thanked the pilot for his extra effort, then shook hands all around. The mystery manila envelope inside my parka poked my ribs—a reminder. I pulled it out, opened the flap: mail, letters, one for me sent six weeks ago, from Annie. It felt warm.

Outside, I viewed blue-sky glory, pure white tops of clouds, soft, puffy, that hid the ice, snow, mountains, beauty, and menace of Marie Byrd Land below us. Our team looked at each other, nodded, and grinned. We were out.

CHAPTER 28

Out-Brief

*If you are a brave man, you will do nothing; if you are fearful,
you may do much, for none but cowards have need
to prove their bravery.*

—**APSLEY CHERRY-GARRARD**, *The Worst Journey in the World*

McMurdo Station, January 15–18, 1990

The Hercules tail ramp dropped open and bright sunlight flooded inside—a light with a quality that I had not seen in a while. Our arrival in McMurdo Monday afternoon felt surreal. We landed at Williams Field, the snow skiway beyond Scott Base. I stepped down the ramp into still air. We dragged our gear over slushy snow. Tucker helped Chris. They each took a strap on a heavy bag and pulled it along. The wet snow sloshed under my steps; the air grabbed me with warmth. This was the start of my vacation. Now this place felt like Club Med. I became happy with relief. Some love resurfaced.

Inside our assigned dorm room, I dumped my pack and Big Red on the floor. I saw a chair, realized I had not sat in one for six weeks, tried it out. Next, I stripped off my layers of dirty clothes and seasoned underwear, dug out my towel, wrapped it around me, and walked down

Tucker and Chris land in McMurdo after Pull-Out.

the hall to the head. I entered the sparse institutional bathroom and looked in the mirror. *I think I recognize him.* Deep tan on his face except for white rims around the eyes, pale elsewhere, thin of body, wild black hair matted in patches, his black beard looked like a worn-out broom. I slipped into the shower and let warm water run through my hair and over my body, places that had not felt a stream in weeks. My hair hadn't had a decent wash for weeks. Then I rubbed soap on me and rinsed off good, realizing I would sleep in a bed soon. Maybe I could do ten hours—that felt good to anticipate.

Weeks without a shower are hard. Camped on the Balchen Glacier, I had dug a snow pit deep up to my shoulders. I heated two pots of water, stripped naked, and climbed down into the snow pit on a calm, sunny day. I stood on an Ensolite pad and had a mountaineer pour over me first one pot to soap up, then a pot to rinse off. Much better than a washcloth rubdown in my tent, but not as good as the real deal in McMurdo.

We went to work, packed our gear and samples for shipment to the continental US, CONUS—home. Chris and Cain organized this. I felt gratitude for Chris' energy and attention to detail. Chris worked hard, got on top of the logistics process. She made our work as smooth as could be imagined. We spent time in the Galley eating decent food. We enjoyed no need to cook and clean up. I hung out in each of the bars.

Word around Mac Town was that the Greenpeace ship, *Rainbow Warrior,* had arrived at Cape Evans to resupply their base that our team had visited on the shakedown trip months earlier. This visit would be my second encounter with Greenpeace. They brought a helicopter, flew south to McMurdo Station, and landed by surprise and without permission on the helipad behind the NSF Chalet. This caused a huge panic in McMurdo and a confrontation with USAP. Greenpeace dared the NSF managers to do anything about them or their helicopter, gave them a big middle finger. They dispatched a crew into the streets of McMurdo to film the trash heap the station presented. NSF alerted all personnel not to speak to them. I missed that memo.

I happened to walk up the road while they filmed, the dusty gravel road between the rear of Building 155 and the three-story dorms—a shabby track lined with dumpsters, supply containers, and telephone poles.

"What's up you guys?" I said to the crew. I figured they had just flown in on that helicopter I saw, but I didn't know at that moment they were Greenpeace.

"We're making a documentary about McMurdo and filming the dumpster here in back of the Galley." I turned to look at the dumpsters. A flock of skuas ripped into them in search of food scraps; they flung papers and garbage onto the road, scattered the debris.

"What do you do here?" a man with a mike said as he shoved it into my face; the other guy started to film me. Cool, maybe they're CNN.

"I'm a scientist, just back from Marie Byrd Land," I said.

"Are you a geologist?" He ignored my answer.

"Yeah." I expected some questions about our research program. Nope.

"What do you think of the ideas for mineral and oil exploration in Antarctica?" he asked. I laughed, told him what a ridiculous challenge that would be, even though maybe both resources were here. After some Q&A, he asked me to sum up how I felt. I'm opposed to resource exploration here, I told him.

"Antarctica is a special place and it's important for us to keep it that way," I said, surprising myself. "Will this be on TV?" I asked.

"Yes, someday soon."

Over a year later, I came home after work and found that my son, Loren, had taped a Discovery Channel TV program on Antarctica. "Dad, you gotta see the end of this TV show," he said, laughed, and fast-forwarded to the ending scene. I stood on that gravel road, closing the TV program, making my brilliant summary statement on camera with McMurdo Sound behind me.

The exposé by Greenpeace and other forces in the US government brought attention to environmental misbehavior at McMurdo by the USAP. Over the years, the situation improved. Today McMurdo still looks like a mining town, but without the trash and debris, without the neglect that NSF allowed in the past.

Friday, January 19

I found Cain in his McMurdo dorm room. Door open, he sat on a chair, backlit before a small window. His thin brown hair lit up in a partial halo.

"Good time for us to talk?" I entered and walked over to him. I felt apprehension—this could be an awkward meeting. "Thought you and I would go over the season, come up with ideas on what worked, what didn't, what to change for next year," I said. He knew I wanted to talk about this. I planned to debrief all the team one-on-one to get their input. I had already spoken to Tucker, gotten his ideas, practical and specific.

I sat on a chair, rotated to face Cain, pulled out my notebook. Cain and I talked about the predictable stuff: amounts and types of gear; what we needed and what we didn't. I paid respectful attention, took notes.

"I won't be comin' back next season with your project," Cain said. "I'm pretty sure the British Antarctic Survey has a slot for me in the Peninsula. That's a long-term job, ya know."

I stopped, taken aback. Not good news. Cain and I had clashed, but he wasn't an amateur. He always knew what he was doing. What a set-back. Cain, our mountaineer, our guide. I took a moment.

"I didn't know your plans, but I see your reasons. Maybe you can help us find a mountaineer."

"Sure, I have ideas. But I have some advice for you, Bruce. You should think about not coming back."

I felt my scalp move, my gut wrench. A nervous laugh burst out.

"Abandon my project? That's not an option. What do you mean?"

"You've got some physical issues." My asthma? He noticed? "Your groin pain laid you up a couple of times. That could've gotten worse, maybe a medevac. You're not strong enough."

My tendon throbbed on cue.

"You don't look comfortable in Antarctica. You're overstressed in this place. You're not good with stressful situations. You worry too much. This isn't for everybody. In Antarctica, nothing's under control."

I had to say something, now way off balance. Once again, I suspected his disdain of my leadership. *Should I push back? Why bother.*

"We got good work done. I made that possible."

"Yeah. The rest of the team seemed to have good ideas too."

"That's a good thing." *I know I'm better than you think. So, fuck off. Be cool, Bruce. Don't lose it.* "Thanks for your input," I said. "I'll think about your advice." I imagined how much nerve it took for him to say this to me. Why did he bother if he wasn't returning with us? I hoped he had sincere intentions, that he wanted it to come across as helpful, not meant as a put-down. I needed to digest his message. I needed time.

There was truth in what he said. But I felt I'd done well enough. Maybe he reasoned that my medical issues could give me a graceful way to exit. Maybe I should stand down. I wasn't proud of how I handled the stress.

I left his room and walked down the dim hallway. My thoughts overcame my sensation of presence. *Who would I be if I didn't return?*

Later in our room, I told Dave some of the practical topics that came up when I talked to Tucker and Cain.

"Hey, Dave, I'm wondering how you feel about next season, if you'll come back, considering what you've been through. I'd never again set foot in this place if I'd survived that crevasse fall."

"I have to come back. I need to finish my work."

"Nobody will think worse of you if you pass." I meant this. "Cain told me he's not comin' back next season. He thinks he's got a job with BAS, wants to keep that gig."

I saw a hint in Dave's face that he already knew this. I didn't tell him what Cain said to me, that I shouldn't return. "That leaves Tucker. He'll probably be available, but I need to ask him."

"I don't want Tucker back. I don't feel safe with him."

"Hey, Tucker rescued you." I flopped on the other bunk, stunned by Dave's remark.

"No, Bruce, I rescued myself." He said this quiet and firm. Dave looked straight at me, to make sure I understood him. I wanted to say, *If Tucker hadn't figured out how to get a rope to you, you'd still be in that crevasse, dead, a solid brick of ice—I'd have made the call to your mother. You guys fucked up, Dave. You untied from your ropes.*

But I couldn't come up with words to speak. Tucker always had input on safety, good ideas and practice, showed me smart caution. I trusted him.

"He's been trying to impress Chris, shows off and loses focus when she's around. He gets careless," Dave said.

"What?" I almost shouted.

"Something happened between them or is happening."

"I don't see that, never noticed anything going on." I didn't probe. I was in a corner, and even more painful, Tucker was my friend. I felt myself in a loyalty bind, a big one. But Dave drew a line.

I studied my notebook to buy time. "I'll go along with you if you tell Tucker yourself the reasons you don't want him back. Tell him you don't feel safe with him."

"Okay, sure, I'll do that."

Cain and Dave had revealed surprises I couldn't have imagined, like Antarctica. Dave not trust Tucker? Would Tucker have told me about Chris? Would Chris have told me about Tucker?

At dinner, I looked for Steve and Chris in the Galley, then sat at their table.

"Cain's not comin' back with us next year," I told them. I could tell by their faces they knew. Maybe Dave told them. Maybe they didn't care.

"Dave doesn't want Tucker to come back either, says he doesn't feel safe with him." I looked for a reaction.

Without hesitation, Chris said, "No, he shouldn't come back."

So, it was true. Something happened between them. Now she wanted him gone. Steve sat silent, busied himself with his dinner. I looked at him; he looked at his food. Did he agree? Did he want to stay out of this? I sat back in my chair, waited—no one spoke.

I decided to tell them something they didn't know.

"I'm thinking I might not come back next season," I said. "My health didn't handle all it had to, not that well. I'm gonna give that some thought." I didn't mention Cain's advice to me.

Chris nodded. "Really, Bruce? Huh. Who'd do your work?" Steve lifted his head to look at me, then at his plate.

Sadness and regret flowed into me. I discovered I worried about and paid too much attention to the wrong things: what the weather would do next; if we'd get our supplies or not; if my asthma would hold us back; if we'd escape this place without getting hurt or killed; if science would come out of all our effort; if I was a fair and effective leader—a leader of a team of equals. That wasn't easy. While I was consumed by all of this, I missed the forest for the trees. I missed the people.

But I made this happen—this extraordinary adventure. None of us would forget it as long as we lived.

CHAPTER 29

The World

Inside of a ring or out, ain't nothing wrong with going down.
It's staying down that's wrong.

—**MUHAMMAD ALI**, World Heavyweight Champion

McMurdo Station, Friday, January 19

"Hey, I got a seat on the Herc north on Saturday," Dave told me. He showed a big grin. We sat on bunks in our McMurdo dorm room.

"How'd that happen? I'm on standby, others too."

"Told them got to get back to my job, to teach, got NSF to speak up for me," Dave said.

"Well, just in time; see you at the MCC." I felt glad he now had a chance to save his job.

"I called my mother from Scott Base," Dave said, changing the topic. "I told her about my crevasse accident."

"You did? Why did you do that? I bet she freaked out."

"Yeah, she kinda did."

"We agreed we'd tell nobody about it. Not NSF or USAP. They could keep us out of the field next year, think we're too careless or it's too risky or something."

"Yeah, I know. I don't think my mom will report us to NSF; do you?" he said with a laugh.

"Did she say she didn't want you to come back here?" I said.

"I'm comin' back, Bruce." I'd give him some time on that decision.

Saturday, January 20

I didn't want Annie to fade away over the months of my disappearance. A phone call from Antarctica would be wise. Christine explained how to do this. "No personal calls from Mac Town, but you can make one from Scott Base—you pay for it," she said.

Scott Base, the New Zealand station, lay about two miles away from McMurdo on the dusty gravel main road over Observation Hill. In November, I had made an appointment for a day in January after our return from the wilderness. I mentioned this in a letter I sent to Annie in December before we left for Marie Byrd Land. We agreed on a date and time—a guess, and a hope on my part that I'd be back from MBL by then.

I walked over the black hill into the face of a painful wind—the hike took about a half hour. A tidy collection of one-story light-green metal buildings, Scott Base stretched up a slope next to the sea ice, isolated from the blight of McMurdo Station. I arrived for my January appointment, walked up the stairs of the main entrance, pulled the lever on a heavy walk-in-freezer door, and stepped into an alcove. Looked civilized—dark blue carpet on the floor, quiet, not the chaos of Building 155 and Highway One in McMurdo.

A young woman with a sun-darkened face greeted me, pointed to my eyes, said, "Just back from the field?" She meant the raccoon suntan we both shared, but unkempt hair and a wild beard framed my features. She led me to a phone booth, stated the cost, that the call would be routed through New Zealand. I closed the wood-framed glass door, took a seat. When had I last sat in a phone booth? Excitement at the idea of speaking to Annie took hold.

I picked up the receiver and it began to place a call. The phone rang in the foreign style of repeated dual beeps. A polite Kiwi operator came on the line. "Call to the States?" she asked, so pleasant. Yes, gave her the number. Anxious, I waited; I had not spoken to Annie in three months. My heart thumped, fingers trembled. I cleared my throat.

I heard the phone ring, with an outer space echo—it rang a few times, and then a click. "This is Annie, sorry I missed your call, please leave a message." I hung up the receiver slow, quiet, dropped my head down, put my elbows on my knees. I stared at my boots. That disappointment hurt.

I persisted. Second time, a day later, she answered. We exchanged chitchat about our upcoming reunion.

"Say, I'll be in Hawaii in a few days. What's your travel plan?" I said.

"Oh Bruce, I'm so tired from work. I don't know if I'm up for traveling."

"Really?" We had planned this trip months back.

"Yes, and other things."

"Like what?" Huh. Maybe a new boyfriend had appeared.

"It's winter here, you know, and I haven't been running. I've gained some weight."

"And so?"

"I don't want to put on a swimsuit in Hawaii where everybody is healthy and trim."

"Huh?"

"I'm just self-conscious, maybe because I'm so tired."

My stomach burned. I was so looking forward to seeing her. Her excuses sounded a bit lame. If you're tired, then why not take a vacation? Overweight? I got stirred thinking about her in a bathing suit—she's just gorgeous. *Okay, be calm and listen, exercise patience.* I wondered if we would go further together. Too early to know, and I was in no position to think about it now.

"Well, okay. I'll contact you from Hawaii and see if you feel better. Love you." We said goodbye.

Next, I got in touch with my son and wished him a happy birthday. I called almost on time, January 17. He turned sixteen. I remember those dismal years in my life. He was doing much better than I had as a teenager. I hoped he knew that I missed him. Hoped he felt special to get a call from me. Maybe he would tell his pals his dad called him from Antarctica. I don't know if he did that.

I packed, suited up in ECW, and showed up for that late-night Saturday flight. At the MCC, my name appeared low down on the standby list. I got bumped, so I said goodbye to Dave, slung my bags over my shoulder, dragged my bag back to my dorm room, and unpacked. That's how I got to stay and see Icestock the next day.

Sunday, January 21

Mid-January is the time for two signature events in Mac Town. One is Scott's Hut Race, a five-mile footrace to Scott Base and back. I watched. Only serious runners enter, but one hundred showed up. Icestock followed that, a garage band concert held on a dirt lot behind the Galley. Our fixed-wing aircraft boss, Rick, helped to organize this.

Located on a gravel yard between the Galley and the Acey-Deucey Club, a bar—a multi-wheel flatbed trailer—made the stage with a parachute as cover for the bands. A couple hundred or so people gathered on the black dust and dirt between the buildings under thin sunlight and a low but cold breeze. Many had cans of beer; others held jugs of what could be coffee. A few danced the best they could in hiking boots and parkas.

Eight groups performed with guitars, drums, and vocals; much of the program showed a lack of rehearsal. The contractors worked six days a week, nine-hour-or-longer days. A thin woman with delicious long blonde hair belted out a song with all her heart. She wore a cutoff t-shirt and jeans—no sign of exposure or frostbite. An original song, "McMurdo Man," held my attention.

> "McMurdo Man,
> You ain't got a plan,
> You ain't got a tan,
> McMurdo Man, you will survive."

This tune got the most cheers. Beer flowed freely.

* * *

I checked the flight manifests posted on the wall outside the Galley. I saw my name and the other four of us listed on a Herc north for Tuesday afternoon—back to the world.

"I'm headed to Honolulu on Wednesday I hope, meeting my girl-friend there," I told the team in the Galley. That might not happen, but I took a chance to announce it. The rest of them had varied plans. Chris had planned a stay in New Zealand. Tucker headed to Santa Barbara; Steve did too. Cain was headed back to Scotland.

Tuesday, January 23

In a Herc once more, we sat packed with others who had finally managed to get a seat north. January is the start of most folks leaving The Ice. By the end of February, the last flight north will leave the winter-over population behind. No flights back south until Winfly in August. VXE-6 recovered the rest of our gear from Depot Camp on Monday, January 20, a week after Pull-Out—late again, but not their fault. A few ANS staffers went to help load. A great adventure for them, to get out of McMurdo virtual jail and fly to the wilderness. Now all our stuff and all of us were out of Mighty Bad Land.

"Look, it's getting dark," I said to Tucker in the web seat next to me. We were over a vacant stretch of the Southern Ocean, a lot of territory.

"Yeah, night, I forgot about that," he said.

Eight hours after takeoff, we landed on a dark Christchurch airport runway, taxied to the entry point for Antarctic arrivals. I stepped down onto the tarmac, my Big Red under my arm, orange bags in hand, and wearing a backpack. *Feels warm here—what's that smell? I know I recognize it.* The smell of green plants, life, and humidity in the air. Unmistakable.

Inside the USAP warehouse, we ditched our ECW, put on our street clothes, and picked up our flight tickets from the staff—not given until all the gear is returned and accounted for.

"My flight to Auckland is soon; I'll be seeing you guys," I said to Tucker, Steve, and Cain. We shook hands, had smiles for each other. Chris wasn't there; she dressed in the women's side of the building. I left, walked out into the warm air, made my way over to the commercial terminal. I never saw or spoke to Cain again. I never tried to apologize for not trusting him, for my unfairness.

Wednesday, January 24

I plunked another New Zealand fifty-cent piece into the pay phone in the Auckland air terminal. It was so loud in there, I couldn't even hear the phone ring. Crowds, noise, advertising—all things I had not experienced for more than two months. Annie got on the line.

"Hey, I'm headed to Hawaii," I told her.

"I'm just wakin' up," she said. I dropped another coin.

"I should be there in about twelve hours," I told her.

"Don't know if I can make it. I'm exhausted from work, not lookin' forward to travel now."

I continued to plunk in coins as we discussed plans. She couldn't decide. I'd be in touch, I told her. *I'm going to Hawaii.*

Now above the Pacific, leveled off at altitude, fasten seatbelts light out, the flight attendant came by and took my order.

"An old-fashioned," I said, the first of a few, my father's drink of choice. I had the row to myself, window on the right side. I looked forward to sleep stretched out for a good part of this eight-hour-plus flight to Honolulu. Out the window, the Pacific spread, visible underneath puffy gold cauliflower clouds, was lit up by a low sun.

Cocktail delivered, I looked at the in-flight entertainment guide. The screen hung on a bulkhead several rows forward of me. I had no control. The crew would start the program when the time came.

While I sipped my drink, I took in the peace of the immense ocean below. I thought of the craziness of our drift through the air thirty thousand feet above. I wrote a journal entry. I had written my first on

November 11, 1989, when we left Santa Barbara. I counted—seventy-seven days ago. Felt like seventy-seven years. What happened? What would I say if asked? In a short time, I would find out that not many questions would be asked. The notion of my experience in Antarctica baffled a lot of folks. Most acted like they couldn't come up with a question.

When the entertainment started and lights went out, I put on head-phones and left my shade up enough to see the fade of sunlight as we flew east into darkness. An ABC Sports segment led off the program. The piece talked about the International Trans-Antarctica Expedition of Will Steger, now underway. He and five companions had made it to the South Pole from the tip of the Antarctic Peninsula. Their final stop would be to reach the Indian Ocean, an equal distance to go.

The studio reporter had a satellite link that enabled him to speak live with Steger. The program showed footage of Steger's team traveling over the white ocean of ice on sledges pulled by huskies. Resupply flights brought back video the team had shot. The coincidence of the event that now unfolded on the aircraft screen startled me. I paid close attention. I gripped my cocktail.

The conversation between the reporter and Steger sounded broken and imperfect. Scenes of snow that blew sideways across the screen and battered his team mesmerized me. Questions were shallow. "What's it like down there?" While I watched, tension built in my chest.

Then, it became difficult to see the screen. Tears. I wept but no sobs. I shot a glance to the adjacent row, embarrassed. No one looked at me. I took my cocktail napkin and blotted my eyes. I watched; more tears appeared. I blotted my eyes once more and my nose began to run. The scenes of hardship crossed the screen one after another as I witnessed Antarctica do what it did best—humble all human efforts.

The segment ended. My eyes cleared and my body felt a rush of warmth that said to me I had been transformed from a state of anxiety to that of peace. At the time, I did not know what my reaction meant and I didn't attempt to understand. But I knew what Steger had done. I understood what he and his team had overcome.

San Diego, February 1990

Home had been San Diego for decades. My mother and aunt brought my two brothers and me here in 1956 after my dad died. I had pestered my mother with the idea to leave the east and move west to California. I was thirteen. I wanted mountains, open spaces, to fish and hunt, and ranches with cowboys. My aunt liked the idea too. My mother sold our home in Oceanside, Long Island.

Our family crossed the country in a new 1956 Ford station wagon she had bought for the purpose—sky blue with white cream seats covered in a clear plastic protector—a V-8 with white-wall tires. When I sat in the shotgun seat, my mother would let me steer with my left hand.

No interstate highways in existence, we drove for three weeks on roads with traffic lights. We stayed in small mom-and-pop motels. Much of the country looked empty, interrupted every few hours by a little town. I saw for the first time the sweep of the landscape when trees of eastern woods didn't block my vision. The sky filled more space than the land. The Rocky Mountains appeared purple at a distance, like in the song. Up close, I found them so high that I stuck my head out the car window to see their snowy tops. Mountain slopes covered dense with trees appeared as dark green velvet like the skirts girls wore at Christmas. This was a different land, I decided, one of dreams that come true.

We crossed the California border in the desert of the Great Basin. Road signs warned of no services until the end of time. We climbed up to scrubby foothills and down the slopes of the semiarid coastal mountains. Soon, we saw people, lots of them in lots of cars.

A cousin lived in San Diego. We drove down for a visit, saw the blue glisten of the Pacific, and stayed. From Long Island to San Diego, the best move that could have been made.

I would forever carry gratitude to my mother for this brave step. I don't remember if I told her that or not. I hope I did. She died on my fourth trip to Antarctica in 1996. I didn't cry when I found out in an email from my brother. I said goodbye at her apartment door and let myself think that I looked at her for the last time. She stood, tiny, gray, and frail, but powerful in the doorway, smiled at me, as I turned and walked away.

The winter after I returned, I had a special visit with her. "My neighbor has letters from Antarctica, old ones, from Admiral Byrd, I think," my mother told me. I sat at the kitchen table in her apartment. This news got my blood pumping. My mother had told her neighbor in the seniors' apartment building where she lived of my recent exploits in Antarctica.

"I told Gladys—she lives across the walk—you'd come back from Antarctica. She said she knew a man who went there with Byrd—he wrote her letters. She still has them." Had my mother been bragging about my exploits? Maybe she was proud of me—could be so. That felt good.

"Your friend has letters from a Byrd expedition, in the thirties? That's incredible." I laughed. "Talk about a coincidence. You know I went to Marie Byrd Land, don't you?"

"Really, you did? I don't understand—all your mail came from McMurdo." My mother saved all my letters from my travels. "I remember the Byrd radio broadcasts from Little America when I was young. It was exciting, so far away, a distraction from the Depression," she said.

"But Mom, we only prepared in McMurdo. We went to the wilderness to work, hundreds of miles away."

"We? Who was that?"

"My team, six of us." I felt a bit exasperated. Why did she forget critical details?

"My goodness, by yourselves? Wasn't that dangerous?"

"Sometimes, but we trained for it." I didn't explain we couldn't train for what we didn't know.

"If I'd known that, I would've worried." She had worried about me on all my adventures since I was nineteen. Thankfully, she waited until I returned to worry. That made leaving her alone much easier on me. But she made her own adventures—she understood what it took—like to haul her three young sons in a station wagon to California.

"Can we go see Gladys now?"

My mother took me to meet Gladys. At her apartment door, a full-figured woman in a white sweater, blue housedress, and arctic white hair greeted us.

"Come in, I've made tea for us." She walked slowly but surely over to her rocking chair, told us to make ourselves comfortable on the sofa. She'd set a teapot and cups on a low table between us. Gladys poured tea from a white pot into flowered cups that sat on saucers. No mugs. We chatted, exchanged forced pleasantries, as my mind obsessed about seeing the letters.

"Your mom says you've come back from Antarctica," Gladys said.

"Yes, a few months ago I went to Marie Byrd Land."

"I've heard about that place. I have some Byrd letters."

"My mother told me. I'd really like to see those."

I noticed a framed vintage photograph on her wall, sepia tone, of a beautiful young woman, nude except for a headscarf, on a beach in a ballerina pose, in profile, a modest version of the arabesque, I believed. Her front leg turned slightly to hide her forbidden forest. Wow. The photo had a Roaring Twenties feeling. I guessed that Gladys saw me glance at it.

"That's me," she said. "My boyfriend was a photographer." I could see why she had a pen pal in Antarctica. Gladys once had been a hot young woman—and a free spirit. I thought I would have liked to know her back then if the clock could be unwound. Next, I felt the sense of loss—of youth, as I contrasted the young beauty in the photo and the very old woman who sat in front of me. I'd seen photos of my mother as a young beauty too—same poignant contrast. *My mom's so frail now. That metamorphosis will happen to me if I'm lucky enough to last long enough.* My Irish grandmother had told me a saying. "Do not resent growing old. Many are denied the privilege."

"A navy man took an interest in me, wrote me a lot at first, from Antarctica and through the war, but I'd gotten married."

She rose to go to her bedroom, came back with six yellowed envelopes. She handed them to me. The first touch of the letters sent an electric thrill through me. Three of them had Byrd markings on them,

"Byrd Second Antarctic Expedition." My hands surprised me; they began to shake a little.

"Wow, Gladys, I can't believe what I'm seeing here," I said. She was silent. I felt her examining me.

"You can have all the letters, the Antarctic ones and others he wrote me during the war," she said. Excitement filled my chest. I couldn't believe these treasures would be mine.

"Really? Gladys, I'll take good care of these. I'll give them to the US National Archives after I read and copy them." I knew that I beamed and couldn't stop smiling.

"Oh, you think they're important?"

"Very important."

"Well, I hope you think so after you read them." She looked pleased with my promise.

Back in my mother's apartment, I sat down at her kitchen table to read the letters with deep attention. My mother sat across from me. I noticed that when I read the markings on the envelopes, my eyes misted—these letters were priceless, historic. Gladys had cut open the envelopes with great care, with a knife or letter opener along one edge, sixty years ago. A precise hand had written these letters on thin, delicate paper. I read them. After a while, my mother said, "What's interesting?"

"These are so special, Mom. Thanks for connecting me with Gladys. I can't believe she gave them to me."

"They must have been important to her; she saved them all this time. She wants them to live on with you."

"I wonder if she remembers what the letters are about. Maybe we should check back and remind her?" I began to read them.

"This guy, Robert English, sent these in 1935. He was captain of one of the two Byrd expedition support ships, the *Bear of Oakland*. I remember reading about that ship," I said. "He wrote her three letters from Antarctica—see? They're chatty, don't reveal personal feelings for Gladys or what he was going through."

"Back in my day, nobody spilled their hearts to anyone, not on paper as I remember."

"I believe it; look at the signatures, 'With old-time best wishes, as ever,' 'Cordially,' and 'With best regards, Always,' no 'Fondly,' 'Love,' or anything that suggested more. I think he played his cards close to his chest, didn't share much of his feelings about his experiences."

"She was married, Bruce. I doubt they're love letters."

"Yeah, in fact they're barely friendly—very odd, like he's trying too hard to be proper, has a hidden agenda."

"This letter is about sailing to Byrd's base at Little America. That's lucky, sailing through the sea ice with no navigation aids, barely a map," I said. "Some bragging in these letters; of course, he earned it. Listen here, he says they had, 'a successful cruise into the unknown regions to the northeast of "Little America," during which we wiped out some 40,000 square-miles of blank space from the chart.'" I smiled and shook my head. "That's a big understatement, but he's trying to take some credit too. A couple of these letters have an amazing history. I can tell from the envelopes. I read about it; I know the story."

"What do you mean? What story?"

"Look what's stamped on the front of the envelopes, in uppercase," I said, handing one to her.

THIS LETTER HAS BEEN DELAYED FOR ONE YEAR
BECAUSE OF DIFFICULTIES IN TRANSPORTATION
AT LITTLE AMERICA, ANTARCTICA

"Delayed a year? Why?"

"Before letters could leave Antarctica, the expedition would cancel, postmark them, 'Little America.' That's what the American public wanted, Byrd Antarctic postmarks for souvenirs. The guy assigned to cancel the letters didn't postmark most of them—there were tens of thousands of letters English brought with him from the US. The un-cancelled letters missed the return of the ship north in February 1934."

"Why didn't he cancel them?"

"Way too many for one guy and he had a nervous breakdown from the workload and tough working conditions. They lived under the surface of the snow, total darkness in the winter." I thought about how grim that would've been. I'd have gone crazy.

"Why didn't Byrd get someone to help him?"

"Byrd didn't know what happened; the worker kept it secret. But when Byrd found out, he asked for a US postal worker to come south the next year and cancel the mountain of mail."

"I don't remember anything about that," she said.

"No, they kept that story quiet. After the ship arrived in January 1935, the relief clerk cancelled the overdue letters and the second new batch of mail that came south with them, thousands. English took all the letters north in February at the end of the expedition."

"So, Gladys got two letters that came one year late," I said. My mother laughed; I chuckled—that was much too cool.

"This third letter has no postmark. But it's got a note inside from an expedition official. He's telling Gladys they'd found the letter hidden in a trunk after the expedition. He apologized for the one-year delay." I showed it to her.

"I wonder if Gladys knows about this," my mom said. "I didn't hear about it at the time."

"This story isn't in Byrd's books. But it got figured out when thousands of people had to wait an extra year for returned letters. I think Byrd omitted facts that embarrassed him. He needed to have a clean heroic story; that's what I think."

"How about that?" she said. "A secret drama. I'll bet there's more. Someone should write a story. If my life had been different, I could write it." The regrets, the loss of opportunity, always near to her surface. "You're lucky, Bruce. Do you know how lucky?" *Is she envious? I can't blame her.*

I reassembled the letters, stared at the stack. Byrd, the legend, confronted what I had. The chaos of Antarctica. I did feel lucky.

Santa Barbara, Friday, February 16

I felt at ease in Andy's office but anxious to begin our therapy session. I brought with me a load of turmoil.

"Back from Antarctica," he said, not a question. He gave me a happy-to-see-you look. All our sessions started that way.

"Tell me about it; I'm interested," he said, tilting back in his Herman Miller chair.

"I didn't do that well. Harder than I expected."

"Why would a stay in Antarctica be easy? That's why only a few people go there." He laughed at his own observation, the irony of mine. I gave a weak snort. He was right, of course. I liked that about Andy—he reacted to what I said, at times with head-back loud laughs, other times with stern messages.

"Tell me more about not doing that well," he said in an even voice.

"I felt frightened often, more than I expected, and anxious." I went on to tell him of storms, cold, whiteouts, hard, hard work, accidents, and the crevasse.

"That sounds completely frightening to me, Bruce. Why would your reaction surprise you?"

"Well, my teammates, they seemed more in control. I'm embarrassed to admit it."

"Do you know what they thought or how they felt? Did they tell you?"
"No."

"So, you don't know, do you." Again, not a question. He scribbled on his yellow notepad. He pulled up a leg to sit on it, leaned forward, to listen with care.

"But the toughest part came at the end, after the crevasse accident. We waited to be picked up, the Pull-Out flight. It kept on getting delayed—weather, schedules. I pretty much lost it."

"Tell me more." I told him.

"After the crevasse accident, I think the weight of responsibility crushed me. I allowed my team to be in danger; I wanted out. But I believed that we'd been forgotten, we would be left there. I sunk into depression." He wrote on the yellow pad, switched to sit on his other leg. That meant he would go deeper with me.

"I'm considering not going back for our second season of work."

He nodded, scribbled. "One of the team said I shouldn't come back. Said I couldn't cut it."

"How'd you feel about his advice?"

"Fuck him. He reminded me nothing's under control in Antarctica. He got that right."

"You couldn't accept not being in control," Andy said. "You were the leader, correct? Accountable for everything?"

"Yeah."

"You've had to be in control your whole life. That started when your mother told you that 'you're the man of the house now,' a few hours after your dad died." He had said this before, but his words sounded new. My gut churned. "That makes a big impression on a child. To get that much responsibility."

"I knew we wouldn't be left out there forever, but I acted like that would happen," I said. "I'm pretty ashamed." *I'm not going to weep. I won't.*

"We've discussed this, Bruce; we've gone over it. Why wouldn't you be afraid of being abandoned? Your dad died, he left you, you were ten. Then two wives left you, then your son. Those are powerful lessons you've learned." This information wasn't new, but I hadn't accepted it before. Now it mattered. Now I got it.

"Remember we talked about loss?" he said. "You're no good with loss—that's being abandoned, that's why you hang on to Annie even though you know different. You know she's not for you."

"How's that?"

"You're afraid to be alone. You've struggled with it almost all your life."

Santa Barbara, August 1990

I entered the geology department mailroom. Lisa sorted the mail, slid pieces into faculty pigeonhole boxes. She looked good, like I last remembered her: dark beauty, smart, funny. We had dated some in the past. I felt good around her, her fun surprises, her humor. I'd told her some of my Antarctic experiences.

"Oh, this is for you." She handed me a large manila envelope. I saw it had a US Antarctic Program label. She saw that too.

"My packet for next year," I said.

"Really? What do you mean next year?" She sounded annoyed. She didn't look at me, continued to fill mail cubbies.

"Yeah, to go back, back to Antarctica."

"Are you crazy? What for? Why? You've got nothing to prove. Let the others take over." She looked at me with an exasperated gaze.

"I'm still deciding—on the fence."

"Oh, maybe I'm interested in what you decide. Let me know." She turned away, kept sorting, stuffed mailboxes. She gave up.

I laughed inside. That was Lisa: blunt. So sweet—maybe she still cared about me. Her observation spooked me. Had she read my mind?

At my desk, I held the USAP envelope in front of me, rested the long edge on the surface, tilted it toward me to read info on the sender. From the medical office—forms to fill out, to start my PQ for the second season. I took a few moments, held that envelope in both hands.

Since I'd returned, I'd learned about what I needed, what I feared, what to fear, and what not to. Antarctica opened a window for me. I came up with an important question I hadn't considered.

Do I want to feel alive?

I sliced open the envelope—spilled out the pile of forms on the desktop. I straightened the edges to make the pages neat. They looked familiar. I worked my chair close to the desk and leaned forward, my head over the pages. I picked up my pen.

I must go back to Antarctica. My job is not finished. I need to be the leader.

Marie Byrd Land, Balchen Glacier; Tuesday, November 20, 1990

Our Herc came to a halt on the Balchen Glacier. I heard the engines power down to idle. The tail ramp whined open; a dim light and sharp cold entered. The Herc lurched forward. The crew slipped out our cargo pallets one by one. I looked out at the snow; it had bright ridges and dark shadows. Our supply line stretched behind us on the snow. All in good shape. The plane stopped. My breath and pulse were under control. Not like last year, that edge, that anxiety.

Our team unbuckled, stood up. I zipped open my parka, dug out my climbing harness from my rucksack. I stepped in the foot loops and cinched it. I watched Dave pull on his harness. He didn't have to come back here. *This is what brave looks like.*

We marched to the rear; the hardware on our belts clanged and chimed, audible even above the engine whine. Our new mountaineers, Terry and J.R., stood at the rear ramp. J.R. tied-in to the aircraft frame, picked up a ski pole, and stepped out. Terry belayed him. J.R. probed along the trail of cargo, probed for crevasses—none found. Great. Relieved—*here goes.*

Dave, Chris, Steve, and I strode down the ramp, stepped on the hard snow. No breeze—lucky—but damn cold still. The team knew what to do. I felt strongly about that—confident. I crouched to set up the radio, Steve the antenna. The rest of the team put up a Scott tent, started a stove, worked with the cargo line. I contacted McMurdo. Now we could stay.

I looked back toward the Herc's interior. The tail ramp raised off the snow, then closed. The pilot applied power. Our team turned away from the blast of snow and wind. The Herc started its run and lifted off. The moan of the engines faded; I watched the plane fly out of sight. *They're gone. We've got this wired.*

I faced south to the front of the Fosdick Mountains. The sun would slip behind them soon—near midnight, I figured. Quiet surrounded us except for the squeak of boots on hard snow, the tinkle of our hardware. I took a long look at the purple profiles of the mountains that darkened with each minute—felt gratitude for their private beauty. Low sun reflected off the hard shine of the rutted snow surface. Chris stepped in front of the light, her shadow busy with our gear. I drew in a breath, fumes from jet exhaust gone into the infinite—pure air now. I let out a slow exhale. Peace settled in me.

We're back. I'm back. We get to discover.

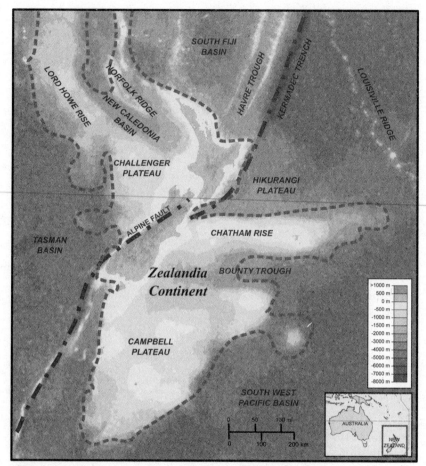

Map 6. *Zealandia today (inside dashed line). Shallow plateaus and rises along with ancient New Zealand were once part of Gondwana thousands of kilometers to the south. The North and South Islands of the nation of New Zealand are now traversed by the Alpine Fault, which formed once Zealandia had moved north. (From Wikimedia Commons)*

CHAPTER 30

Finding Zealandia

Adventure may hurt you, but monotony might kill you.

—**APRIL VOKEY**, Fly Fishing Guide

Santa Barbara, 1994

"Bruce, you know John Bradshaw is visiting here." Chris had stopped by my office in the geology building at UCSB. "He's agreed to give a few of us a short talk about some of his new findings on New Zealand geology."

Later that day, we gathered in my lab and sat around a wide, black-topped worktable. John was an OAE, a veteran of MBL, and a respected geologist at the University of Canterbury in New Zealand. For me, the main take-away from his talk was convincing evidence that the island nation underwent a very sudden tectonic change from a time of compression in a subduction zone—where an oceanic tectonic plate is sliding under an overlying plate on its way to the Earth's mantle—to suddenly stretching and rifting 105 million years ago—a reversal of tectonic forces. It happened in a blink of geologic time; that was unusual. He ended with a question: what caused this sudden shift in tectonic forces?

His evidence for a sudden change in the tectonics at that time, and asking what caused this, thunderstruck me. I thought I could answer his question. I had the rare, elated feeling of discovery that makes research so addictive—I felt the rush. I stopped Chris on our way out of the lecture.

"Hey, Chris. Our findings in the Fosdicks seem parallel with what Bradshaw just talked about for New Zealand," I said.

"Maybe, but like what? The timing?" she said.

"Yes, for one. We found a stretching event in the Fosdicks that also started at about 105, the same timing Bradshaw found for New Zealand. It was the start of uplift of the Fosdick Mountains. Also, the metamorphic rocks of the Fosdicks and metamorphics in New Zealand are relatives. They're the same age and they fit close to each other when Gondwana is put back together—you know this. The mechanism and the timing in New Zealand and MBL could be related to a common process. I have some ideas."

"Ready to share?"

"It's about California. I need to think it through." She gave me a smile and a nod.

I was finishing work on a story for California that gave me an insight on a possible process in common. My colleagues at UCSB and I had proposed a theory to explain why tectonic compression recorded in Southern California geology had suddenly changed to stretching about twenty million years ago. We proposed that an oceanic plate subducting under California reversed direction and started pulling away from California instead of sliding under it. It fragmented the overlying western edge of North America and broke parts of it away. What if the Gondwana breakup was like what we proposed for California—a plate subducting under Gondwana reversed direction and started pulling away from the supercontinent instead of sliding under it? The plate would break off chunks of Gondwana and carry them away northward, like pulling the rug out from underneath it. It started 105 million years ago and should show in the geologic record of both MBL and New Zealand—just like we see.

Now it was my job to weave a story, an explanation for the dispersal of pieces broken off the Gondwana continent. Not all necessary data existed, but it was enough to do that. I studied a research paper published two years earlier by scientists at the University of Texas. That paper had a crucial map model. It showed the sunken plateaus, now scattered around the Southern Ocean and New Zealand itself, attached to MBL in Gondwana back at the time before rifting and drifting began. It was a very neat puzzle fit that had not been obvious at first. But the map model clearly implied that all the pieces were parts of a continent—Gondwana.

Before 105 million years ago, the Phoenix oceanic plate was subducting under the continent of Gondwana. The now sunken plateaus along with what would be New Zealand were still part of that continent. The Phoenix plate was trailed by the Pacific oceanic plate. As the Phoenix plate diminished in size due to disappearing into the mantle under Gondwana, the Pacific plate captured the smaller Phoenix plate and started dragging it north, away from Gondwana, which was above Phoenix in the subduction zone. Traction between Gondwana and the underlying captured plate—the rug underneath—broke off overlying pieces of the continent, the one-day sunken plateaus and New Zealand, to travel with it. The region of Gondwana left behind included what would become modern Marie Byrd Land.

Later, other plate tectonics distributed the rifted parts of Gondwana into errant pieces, separated and sunken below the waves. The pulling away by the Pacific plate had stretched and thinned the captured Gondwana fragments, so they sunk—except for New Zealand, which remained mostly intact, but not of its present shape.

As I looked at the Texas model and at the modern-day map of the Southern Ocean and its plateaus, along with New Zealand sitting atop two of them, I realized that a simple label was needed to contain my hypothesis. All these distributed pieces were continental—from Gondwana, and all were once joined within it. They are the closest relatives. New Zealand is the most prominent because it's above the ocean surface, but it's only a fraction of what lies below sea level.

A name to knit the dispersed continental chunks together as an idea was needed and that name had to include New Zealand—Zealandia.

Santa Barbara, 2012–2014

I published my story in 1995 in the scientific journal *Geology*. The reaction from geologists was muted, most of them concerned about the plate tectonics I had proposed, which was speculative. Then, attention shifted to other implications. In 2012, I received an email from a New Zealand colleague, Nick Mortimer. He wanted to know about the name Zealandia.

> *Hamish Campbell and I have been asked by Penguin to write a book on "Zealandia"…can you give us any background as to how and why you came up with the word Zealandia?*

My response was:

> *I was trying to think of a good way to name the inclusive areas of NZ, Campbell Plateau, Lord Howe, etc., in order to convey a geographic and geological relation to each other; New Zealand itself was the centerpiece. "Land" had to be in the name and "Zealand" was taken, so I must have thought of an easy way out: "Zealandia."*

Two years later, in 2014, they came out with their book, *Zealandia: Our Continent Revealed,* a masterful volume written in a popular science style with numerous color photos, including one of me that noted I originated the name and concept. It was a comprehensive book covering "its identity, discovery, origins, separation [from Gondwana], life, submergence, emergence, and resources." The last word in that description held a key fact: New Zealand could now claim vast areas of submerged territory as its economic property.

Santa Barbara, 2017

I received a surprise email from Dave Mosher, a reporter at *Business Insider*. He was writing a news article: "Earth has a brand-new continent called Zealandia, and it's been hiding in plain sight for ages." A research paper had just been published in *GSA Today* by Nick Mortimer and a group of New Zealand geologists that added scientific teeth to the Zealandia hypothesis. Mosher said I was credited in that research paper; what did I think of this? He wanted quotes and my response to the claims in the research paper that Zealandia is Earth's eighth continent. I explained that the submerged pieces once were attached to Gondwana, which was a supercontinent, but looking at the *GSA Today* research paper, I agreed with the authors' criteria. The submerged plateaus are continental because they meet the criteria, as Mosher paraphrased in *Business Insider*:

1. Land that pokes up relatively high from the ocean floor.

2. A diversity of three types of rocks: igneous (spewed by volcanoes), metamorphic (altered by heat/pressure), and sedimentary (made by erosion).

3. A thicker, less dense section of crust compared with surrounding ocean floor.

4. Well-defined limits around a large enough area to be considered a continent rather than a microcontinent or continental fragment.

I explained to Mosher that people tend to think of continents as places to live that are above sea level, but geologists use the four points stated in the article to identify continental regions. Sea level changes drastically over time—some hundreds of meters when ice sheets appear and disappear, for example. Continents are stable even if they drift around on plates while the ocean floods and drains off them.

Many press interviews followed the publication the *GSA Today* paper, with me, and more with Mortimer and Campbell, plus other authors of the research paper. Most of the press focused on the lost continent metaphor. How big is it? As the *GSA Today* article had pointed out, it's "approximately the area of greater India" and thus can be thought of as the eighth continent. The New Zealand nation rides on two of the sunken pieces and is the only significant location where Zealandia pokes above sea level; but still, that nation represents only a few percent of the area of sunken Zealandia.

Besides these geologic facts and arguments, I think my most significant quote published in *Business Insider* was this: "The economic implications [of Zealandia] are clear and come into play: What's part of New Zealand, and what's not part of New Zealand?"

Dave Mosher explained the significance of this question in his article: "United Nations agreements use continental margins to determine which nations can extract off-shore resources—and New Zealand may have tens of billions of dollars' worth of fossil fuels and minerals lurking off its shores."

Epilogue

What do I remember now, looking back across the gap of
so many years to those moments of great living in a distant land?
I remember the crevasses, the wind, the great white slopes....
I remember the silence. But I remember most that sense of peace,
transcending human care and the violence of the wind,
that reigns in those lonely places.

—ROBERT DODSON, *Crevasse*

Santa Barbara, 2016

The incoming call alert on my cell showed Colorado—Christine. Hadn't spoken to her in a year or more.

She was now a professor of geology at Colorado College and continued her Antarctic work. Dave had a successful career underway teaching and conducting research at San Diego State. When he returned from our expeditions, he married and raised two daughters—one became a geologist. Steve married, raised two daughters, and had a career at the Arizona Geological Survey, where he retired. He continued consulting in management of Earth and environmental data in a project sponsored by the NSF. Tucker married and moved to Portland, Oregon, where he raised two daughters and developed an environmental consulting business. Sadly, he died in October 2022. Cain went back to Scotland and the French Alps, where he continued to guide climbing expeditions. He spent his summers in Scotland, exploring it by bicycle with his woman partner, leading cycle tours, and had published a book in 2015, *Scottish Cycle Routes*.

Me, I visited Antarctica eight more times, the last in 2010–2011. Of these visits, two were on-continent expeditions, like the story in this book. Three were marine expeditions on research icebreakers, and the rest on-continent, but associated with complex geophysical experiments. On one trip in 1998, I visited my son, Loren, mentioned in this book, in McMurdo. He was working for Antarctic staff support services managing the hydroponic greenhouse. I retired from teaching in 2010. During this period, I met Susan, and we married.

I answered the Colorado call from Chris; my voice revealed I was happy to hear from her. "Hi, Chris. How are you? What's up?"

"Hi, Bruce. I hope this a good time. I have great news for you." She spoke in her deliberate and careful way. I could tell that she had something important to say.

"Always open for great news."

"The US Board on Geographic Names just approved naming a mountain in Antarctica after you."

"What? Really? How come?"

"Chris Sorlien and I nominated your name for peak 1070 in the Fosdicks. It's now Mount Luyendyk."

"Wow! 1070 above our Birchall Camp? Where we had spent Christmas and your birthday and that ridiculous storm? It's so beautiful. I'm speechless." I was choking up.

"Listen, Bruce, I'll read part of the citation."

"Named for Professor Bruce P. Luyendyk, who has been active in international Antarctic research for 25 years. He was responsible for two onshore expeditions and five marine geophysical expeditions in the Ross Sea. His cumulative research, findings, and publications have significantly increased scientific knowledge in Antarctica."

"This is so great. Thank you, Chris. Thank you for doing this," I said.

Epilogue

"You've earned it. There'll be a presentation ceremony at UCSB soon. I'll come out to Santa Barbara. The entire geology department will be there. I have a terrific photo of your mountain. Took it in 2010 during a fly-in. I'll email it to you."

After I hung up I sat there in wonder. How to let her news sink in? I was almost dizzy with joy. Chris and I collaborated since our early expeditions and she led several returns to the Fosdick Mountains with her own teams; I began to call her Queen of Marie Byrd Land. I thought back on the body of work we produced, starting with the two FORCE expeditions in 1989–1991.

We figured out how these ranges formed and why. The biggest surprise was that age-dating and study of the metamorphic minerals showed the ranges arose before the rifting away of New Zealand and its continental partners—some twenty million years before sea floor spreading began at eighty-five million years ago. The Fosdick Metamorphic Rocks formed at this time also, more than one hundred million years ago. Rather than continental rifting as a cause, it was subduction of the Phoenix oceanic plate under the edge of Gondwana that triggered the metamorphism and mountain building. A great deal of heat was produced under MBL during subduction and it stuck around for millions of years, which partly explained the fact that the Antarctic side remained higher than the New Zealand side after rifting began. The other reason for the asymmetry of rifting was stretching and sinking of Zealandia pieces as they were being pulled away from Gondwana when the Pacific plate captured the Phoenix plate. My own paleomagnetic work showed that the MBL ranges tilted and twisted clockwise as a whole during and after the mountain building and helped connect the metamorphism with that.

Careful age-dating work by Chris provided evidence of the rocks that preceded the Fosdick migmatites—the rocks that underwent metamorphism to make the rocks of the Fosdick range. These include a widespread sedimentary unit known as the Swanson Formation, some 450 million years in age, that spans parts of East Antarctica and Australia. Along with that rock, the Ford Granodiorite we encountered in the

Chester and Phillips Mountains, about 360 million years old, was also metamorphosed to form the Fosdick Metamorphic Rocks. We found evidence of that process at Neptune Nunataks (chapter 14).

The magnetic strain data, or AMS measurements, revealed stretching deformation that had not been oriented at right angles to the Fosdick range as would be expected, but oblique—more northeast than north, or clockwise deflected. That suggested the present trend of the mountains was established at some time after the metamorphic rocks were deformed. It means there could have been two tectonic episodes.

The observation of glacial grooves we made on Swarm Peak on Christmas Eve in 1989 was not forgotten (chapter 19). We teamed with a glacial geology expert who accompanied Chris on a subsequent expedition. Observations and age-dating of glaciated rock surfaces put together a story of ice sheet retreat since seven thousand years ago. That would be explained by the global interglacial period Earth has entered. But it started much later in Antarctica than in the Northern Hemisphere. Today, that retreat continues and is accelerated by global warming.

I opened her email and looked at Chris' photo of Mount Luyendyk, taken from the air at low altitude. The view to the southeast exposed the tall gray cliffs of Mount Luyendyk. Snow capped the peak and blanketed the south flank. A lake of blue ice stretched in front of the mountain on the Balchen Glacier. Happiness wrapped around me. I tried to find our 1989 campsite in the photo. A bit out of the frame, I guessed. What would that spot look like now? What was happening there at this moment— that place where we endured a ferocious blizzard? The raw, penetrating danger and thrill of surviving that storm surfaced in me. The impossible aloneness of that place came back into my mind. I thought of my privilege to experience Antarctica in its beauty and harshness and accepted gratitude. I let myself dream of the white ocean of ice.

Appendix

Author Notes

A few basic facts will help the reader stay oriented in this unique and unusual place. To begin, why are parts of Antarctica named East or West Antarctica? In the map illustrations, West Antarctica is shown east of East Antarctica, presenting an immediate apparent contradiction. The naming convention is tied to what parts of the continent are in east versus west longitudes (see Map 1). What are longitudes? These are the map meridians or lines that run from pole-to-pole on the Earth. The prime meridian is zero degrees longitude and runs pole-to-pole through the Royal Observatory in Greenwich England. On the opposite side of the Earth, the antimeridian, 180 degrees of longitude, runs from pole-to-pole through the Ross Sea of Antarctica. These meridians divide the globe into longitudes east and west of Greenwich. The part of Antarctica east of 180 degrees is in west longitudes, hence West Antarctica; the opposite holds for east longitudes and East Antarctica. A geologic contrast also applies to the East-West division; East Antarctica is higher altitude, larger, and geologically older than West Antarctica. A clear geologic boundary is the Transantarctic Mountains that are on the edge of East Antarctica and border the Ross Sea and Ice Shelf (these are mostly in geological West Antarctica).

The use of common geographic cardinal directions north, south, east, and west can also get confusing in Antarctica. After all, standing at the South Pole, all directions are north. Elsewhere, away from the

Pole, in Marie Byrd Land for instance, north, south, east, and west can be applied in a local area. We used our Brunton geologic compasses to find the local cardinal directions. However, we were located so close to the south magnetic pole that our compass needles aligned with geographic east-west, not north-south. I determined this deviation for the area where we worked and we adjusted our compass dials to account for this offset (chapter 13).

The convergence of meridians at the Pole makes air navigation with the common global latitude-longitude lines of parallels and meridians problematic. Navigators use a polar grid system where the prime meridian is grid north, 180 degrees longitude is called grid south, 90 degrees east longitude is grid east, and 90 degrees west longitude is grid west. In the story, conversations by radio between the expedition team and aviators are using grid directions (chapters 13, 20, 27).

Also, in the illustrations and parts of the story, the metric system is often used. Distances are in places given in kilometers. An easy conversion to remember is ten kilometers equals six miles; this is labeled on maps. Elevations are in meters (3.28 feet per meter), where one thousand meters is one kilometer. Temperatures are given in degrees Celsius where zero degrees is the freezing temperature of water (32° for Fahrenheit), and zero degrees Fahrenheit is minus 17.8 degrees Celsius. Wind speed in the story is typically stated in knots; for a conversion example, a speed of 10 knots is 11.5 miles per hour.

Time of day is also unusual on a continent that spans all the time zones on the globe. New Zealand time is used for McMurdo local time and all operations based out of McMurdo. Therefore, our expedition used McMurdo-New Zealand time and a twenty-four-hour clock. Scientific observations, on the other hand, are typically logged to time at Greenwich on the prime meridian. This time is called Coordinated Universal Time or UTC (and sometimes called Zulu time; see chapter 23) and applies everywhere on Earth, day or night. For instance, if we made weather observations at 0700 (7 a.m.) McMurdo-New Zealand (standard) time, it would be logged as 1900 (7 p.m.) UTC the previous day (minus 12 hours).

A reader might get the impression from the story that Antarctica is US territory—at least Marie Byrd Land. Portions of Antarctica are claimed by nations, but the international Antarctic Treaty System has suspended these claims. The US has made no claims. Thus, no nation has authority or sovereignty over any parts of the continent. The US has established dominant bases across the continent—at McMurdo Station—which is part of this story, at the Geographic South Pole, and on Anvers Island on the Antarctic Peninsula (Palmer Station). The McMurdo and South Pole stations were first established during the International Geophysical Year (1957–1958), with Palmer established between 1965–1968, but now are regulated in a cooperative sense by the Antarctic Treaty System if solely operated by the US.

Acronyms

The US Antarctic Program is a vestige of the original US military exploration in Antarctica beginning in the mid-twentieth century. Following on from this, the civilian science program adopted many of the convenient and systemic uses of acronyms in the US military. In alphabetical order, with the chapter where it is first introduced:

- **AMS** (Chapter 23). Anisotropy of magnetic susceptibility. Many rocks contain magnetic minerals, a portion of which are needle-shaped. Under tectonic pressure, these needles can be aligned by the external forces. This alignment in a rock sample can be measured with an instrument that senses the difference in magnetic response in various directions. These types of data are useful in detecting tectonic events that deformed the rock and the formation.
- **ANS** (Chapter 4). ITT Antarctic Services. Government contractors that manage the infrastructure of the Antarctic stations such as McMurdo and South Pole.
- **BFC** (Chapter 5). Berg Field Center. A facility in McMurdo where field parties obtain tents, climbing gear, and food rations.

- **CDC** (Chapter 1). Clothing Distribution Center. A US facility in Christchurch, New Zealand, where those headed south to The Ice get their cold weather clothing.
- **CNN** (Chapter 28). Cable News (television) Network. A US network with global distribution.
- **CONUS** (Chapter 28). Continental United States. A jargon for shipping destination.
- **ECW** (Chapter 1). Emergency Cold Weather clothing issued at the CDC.
- **FDM** (On some maps). The Fosdick Mountains. Used on cargo labels for items headed into the field.
- **FNG** (Chapter 1). Fucking New Guy. A holdover from military jargon identifying new Antarctic program participants.
- **FORCE** (Chapter 4). An acronym to identify our project, Ford Ranges Crustal Exploration.
- **KIT** (Chapter 10). Keep It Together. Said when things got rough.
- **MBL** (Chapter 1). Marie Byrd Land, a region on the Pacific side of West Antarctica and the location of the story. Its scattered mountain ranges are distributed over an area about the size of California.
- **MCC** (Chapter 17). Movement Control Center where passengers headed north, gathered before boarding aircraft, and where mail call was held.
- **MOGAS** (Chapter 13). Motor gasoline, or ordinary fuel for gasoline engines. Used to distinguish it from jet fuel.
- **NCO** (Chapter 5). In the US military, a noncommissioned officer, a senior rank in the enlisted corps as opposed to the officer ranks.
- **OAE** (Chapter 1). Old Antarctic Explorer, or a veteran who's been to The Ice before.
- **PI** (Chapter 1). Principal Investigator. The lead scientist of a project.
- **PNR** (Chapter 1). Point of No Return. The halfway point between Christchurch and McMurdo beyond which an aircraft does not have enough fuel to return to its origin.

- **PQ, NPQ** (Chapter 2). Physically Qualified, Not Physically Qualified. Results of medical and dental exams necessary to qualify or not for Antarctic service.
- **SAR** (Chapter 5). Search and Rescue. Used both to describe the process and to name a specialized rescue team.
- **UCSB** (Chapter 5). University of California, Santa Barbara; the host university for the FORCE project participants.
- **USAP** (Chapter 1). United States Antarctic Program. The federal umbrella organization that governs all US activity in Antarctica.
- **USNSF, NSF** (Chapter 1). The National Science Foundation, the United States federal agency that funds most American basic research, including that in Antarctica.
- **VXE-6** (Chapter 4). Antarctic Development Squadron Six, a US Navy air squadron based at Point Mugu, California, Christchurch, and McMurdo. It performed the primary air support for the USAP until 1999 when that task was assigned to the US Air National Guard.
- **WAVE** (Chapter 5). West Antarctica Volcano Expedition. The research team deployed in far-east Marie Byrd Land when the FORCE team was also in MBL.

Every Rock Has a Story to Tell

This book is a story about people who are geologists. They were seeking the history of Gondwana and Zealandia by looking for the stories captured in rocks. A useful concept for the reader not familiar with geologic terms is to understand that geologists are studying rocks that are parts of a formation, not miscellaneous cobbles that might be found in varied locations. A formation is an expanse of the same kind of rock that spans an area large enough to be shown on a map. For example, the Fosdick Metamorphic Rocks is a formation that comprises the unique rock type of the Fosdick Mountains. It spans about thirty kilometers above the ice surface and maybe more below that. Here are the main formations and a brief description of their characteristics. Also included below are some significant minerals that are found in the rocks of these formations and are mentioned in the book:

- **Fosdick Metamorphic Rock.** A metamorphic rock formation— one formed by alteration of a preexisting rock by heat and/or pressure—that makes up the Fosdick Mountains that we studied. The metamorphism occurred more than 100 million years ago by the alteration of the Swanson Formation, a sedimentary rock formation in Marie Byrd Land that is dated at 450 million years or older.

- **Gneiss.** A metamorphic rock type with visible layered banding of light and dark colored minerals. Common in Fosdick Metamorphic Rocks.

- **Migmatite.** A metamorphic rock type with contorted bands of light and dark minerals. It has the appearance of having been partly molten and then crystallized. Common in Fosdick Metamorphic Rocks.

- **Byrd Coast Granite.** A formation about 100 million years old spanning the area of the expedition and mountains south of our study area comprised of granite, a common igneous rock (crystallized from a molten state deep in the crust). It is mainly comprised of the minerals quartz and feldspar. Not present in the Fosdick Mountains.

- **Ford Granodiorite.** A formation found in the Phillips and Chester Mountains and areas south. Granodiorite is an igneous rock like granite but with more dark minerals, hence its dark gray-purple appearance. The Ford Granodiorite is older than the Fosdick metamorphics and Byrd Coast formations—about 360 million years. Some of it has been metamorphized to produce the Fosdick Metamorphic Rocks.

Volcanic Rocks Mentioned

Volcanic rocks cool and solidify from molten magma at the surface of the Earth or close to it. The variety of volcanic rock in our study area is basalt. This is a dark gray to black rock devoid of lighter-colored minerals like those that make up granite, for instance. Two occurrences of volcanic rock are mentioned in the story. A small inactive volcano, Mount Perkins, is located on the eastern Balchen Glacier. It is dated at 1.4 million years. At Avers Camp on the Balchen, there is a pile of volcanic cinders the size of a large house (chapter 23). Because the glacier has not eroded it, the cinder cone might be only a few thousand years old.

Dikes

Igneous dikes of basalt are common and first mentioned during our visit to the Chester Mountains (chapter 13). Dikes form from magma that fills sheet-like cracks or fractures in older rocks and then solidifies. They are an indicator of the crust having been stretched and cracked to create space for dikes to invade formations. They can appear as dark stripes across an area or as low walls where the older rock has eroded away. Dikes occur in all the formations of the research area and are 100 million years or older.

Key Minerals Mentioned

- When geologists say they have identified or noted specific minerals in a rock specimen, think of crystals, usually minute and best visible with magnification. The main minerals in the Byrd Coast granite are glassy quartz and tan orthoclase feldspar. The Ford Granodiorite has less quartz but also gray-purple plagioclase, a different variety of feldspar than found in granite.
- Biotite is a black mica mineral that makes up the dark layers and bands in the Fosdick gneiss and migmatite. The lighter bands are quartz and feldspar.
- Key minerals that are found in small amounts in the Fosdick metamorphics are cordierite, garnet, and sillimanite (chapter 19). Garnet is the familiar gemstone and cordierite is the less common gem iolite. Sillimanite is not a gem mineral, but it is used in manufacture of some types of glass. These minerals are

fingerprints of the conditions of heat and pressure that the Fosdick Metamorphic Rocks endured.

- Also found in small amounts are the minerals monazite and zircon (chapter 16)—the latter is also a gemstone. These minerals are very robust and can survive tens or hundreds of millions of years. They contain small amounts of radioactive uranium that decays over time to lead at a regular rate. Measuring the content of uranium and lead in these minerals can yield a maximum age date of the creation of the mineral and thus the rock and formation that contains them.

Scientific Findings

By Bruce Luyendyk and Christine Smith Siddoway

The main thrust of the FORCE expeditions of 1989–1991 included determining the history of the Fosdick Mountains, a unique massif in the northern Ford Ranges of MBL, thought to have a history correlative with similar ranges found in New Zealand (Tulloch and Kimbrough 1989, Richard et al. 1994), the former partner to the region we studied. We tested the hypothesis that the metamorphic rocks of the Fosdick Mountains, largely migmatite and paragneiss, are the central exposures of a metamorphic core complex like those well known in the Cordillera of the United States (Coney 1980). We benefited from the prior reconnaissance geologic mapping of the region in 1966–1967 by Wade, Cathey, and Oldham (1977). The Fosdick Metamorphic Rocks were first described by Wilbanks (1972).

Metamorphic rocks are those formed by heat and pressure applied to another rock, the protolith, such as one composed of sediments (sedimentary) or a rock crystallized from the molten state (igneous), or even from a metamorphic protolith. Migmatite literally means mixed, and in this usage, it refers to mixed metamorphic and igneous rock appearing together in distorted bands—a marble cake is a good analogy. A gneiss is a metamorphic rock formed from an igneous or sedimentary rock and shows thin sheets of both dark and light-colored minerals. The para designation (e.g. paragneiss) indicates the protolith of a particular metamorphic rock was sedimentary rock.

Metamorphic core complexes are created from extreme crustal extension absorbed in flat-lying faults known as detachment faults (Davis and Lister 1988). The crust is stretched to such an extent that upper layers are pulled away to promote the upwelling of deep metamorphic rocks into the gap or core. We first thought that the rifting away of the New Zealand microcontinent from this portion of West Antarctica about 85 million years ago created the needed extension that allowed deep metamorphic rocks of the Fosdick Mountains to reach the surface. The facts proved more complicated.

The major results of the FORCE project are reported in publications that appeared from 1992 to 2011. Our detailed field mapping at first failed to find signature structures of a core complex, such as a detachment fault. Moreover, the timing of events we found from studies of the cooling history were earlier than 85 million years (Siddoway et al. 2004, Richard et al. 1994). What then caused these mountains to be exhumed—to rise to the surface in a manner of speaking? The answer took a few years to come into focus.

Steve Richard, who named the mystery mineral "blueite" in chapter 19, "Swarm Peak," determined later to be cordierite, took the lead in unraveling a cooling history of the Fosdick Metamorphic Rocks (Richard et al. 1994). The cooling experiments have the engaging name of *thermochronology*. The method seeks to determine when specific minerals cooled below a blocking temperature. Below this temperature, radioactive decay of elements in the minerals begins. Thus, the results are a time-temperature relationship for the rocks containing these minerals. The uranium-lead decay starts in the mineral monazite at about 730°C. Dave Kimbrough determined these ages. Dave's specialty included field geology and lab studies of isotopes in rock minerals. He used mass spectrometers to determine ages using the radioactive decay series of uranium to lead. Another method uses isotopes of the gas argon trapped in the minerals to determine when the clocks were set. Steve Richard did that work. The last method used looks at tracks made by radioactive decay products in the mineral apatite. Tracks are not made until apatite cools below 120 degrees Celsius. With the passage of time, more tracks accumulate. Our colleague Paul Fitzgerald, now at Syracuse University, did these experiments.

What did we find? Cooling of the Fosdick rocks began about 105 million years ago when the rocks were at 725 to 780°C. The metamorphic minerals found suggest the rocks were more than 18 kilometers deep at that time. From 101 to 95 million years, cooling was fast—around 70°C per million years. This cooling rate suggests rapid exposure (coming to the surface). Following 94 million years, cooling was much slower and had interruptions that infer the region responded to the breakaway of New Zealand and Zealandia from Gondwana during the later period. The important fact is that the metamorphism and exhumation of the rocks of the Fosdick Mountains began well before the breakaway of the New Zealand and its microcontinent partners.

A few years after FORCE, Christine Siddoway led a project (Siddoway et al. 2004) that reexamined much of this early work and incorporated results from airborne geophysical surveys over the Ford Ranges on which we collaborated (Luyendyk, Wilson, and Siddoway 2003).

The airborne work comprised flying a small aircraft outfitted with instruments to measure the Earth's gravity and magnetic field, along with ice thickness, on a grid of lines over the Ford Ranges. The flights were coupled with on-ground field mapping of rock types and fault networks by Chris and colleagues. Collating the airborne and ground data, they mapped buried features and faults. One conclusion from the research was that the northern Ford Ranges lay within the eastern boundary of the West Antarctic Rift province (Behrendt et al. 1991). The province boundary is well located in the west up against the Transantarctic Mountains, but the east side wasn't mapped until we surveyed it with gravity, magnetic, and geologic data (Luyendyk, Wilson, and Siddoway 2003).

For the new research on the metamorphic rocks, Chris brought on board Mark Fanning from Australia, who Chris met during a field excursion in Japan. He provided Chris with facilities for detailed age-dating. One aim was to use ages of the mineral zircon to determine the original rocks, the protolith, that preceded the Fosdick Metamorphic Rocks. The question asked was: what rocks underwent such extreme heat and temperatures to become the Fosdick migmatite gneiss? We already realized that the nature of the Fosdick rocks themselves indicated that they formed from clay-rich sedimentary rock. The prime candidate was the Swanson Formation, of Ordovician age (450 million years), found tens of kilometers south of the Fosdicks but also in Northern Victoria Land of East Antarctica and eastern Australia—it's a pan-Gondwana formation.

Fanning provided Chris with access to an instrument called a SHRIMP (sensitive high-resolution ion microprobe) to determine ages of zircon grains extracted from Fosdick rocks. Zircon is a highly robust mineral, and it survives intense metamorphism, so the idea was to look for ages that matched ages of zircons found in the Swanson. And that's what she discovered—zircons of early Paleozoic age like the ones found in the Swanson Formation (~560 to 520 million years and some of Proterozoic age, as old as a billion years). Chris also found zircons with an age that matched ages of zircons from the Ford Granodiorite—Devonian, suggesting that these rocks also were metamorphosed to form parts of the Fosdicks. The Ford Granodiorite is found north of the Fosdick range in the Phillips Mountains and south of it in the Chesters. With these results, we had a good idea of what rocks were metamorphosed to form the Fosdick rocks. But when and how?

Next, Chris determined the pressure (depth) conditions the Fosdick rocks endured. She did this by using pairs and sets of metamorphic minerals (Siddoway et al. 2004). Later, with collaborators, she used sophisticated modeling methods to refine the estimates (Korhonen et al. 2011, Korhonen et al. 2010). By merging these results with the time and temperature data, a story took shape, a story refined from the original history, but the same in important ways. We defined four deformation events over 430 to 98 million years. The most intense metamorphism took place between 136 and 116 million years at a depth of more than 18 kilometers (600 megapascal)—when a temperature of over 700°C was reached. After this, the rocks rose to shallower depths (near 10 kilometers), and the mineral cordierite formed (blueite!). At 98 million years, rapid cooling at 70°C per million years began.

Scientific Findings

We concluded that a major fault had to run along the Balchen Glacier at the north front of the Fosdick Mountains (Siddoway et al. 2004, Luyendyk et al. 1992, Richard et al. 1994). Not only did the topography and contrast in rock formations across the glacier suggest that, but the cooling ages showed that the Phillips Mountains on the north side of the glacier cooled well before the Fosdicks. One way to explain that would be if a fault ran along the south edge of the glacier in front of the Fosdicks—thus, the name Balchen Glacier fault. This fault had to accommodate the rise of the Fosdicks. This conclusion relieved us of the quest for a detachment fault to explain the rise of the Fosdick complex. In spite of several field seasons of mapping, a detachment evaded detection—if it in fact existed.

The other data brought to bear on the story included strain indicators—the style and direction of deformation. One set of data included the magnetic strain data, the anisotropy of magnetic susceptibility, or AMS, measurements discussed in the book. A key result was that the strain was not oriented at right angles to the Fosdick range and Balchen fault as expected, but oblique—more northeast than north, or clockwise deflected. Consistent with this, the northern Ford Ranges themselves trended oblique to the continental margin—not parallel to it as might be expected if their origin was related to the rifting away of New Zealand and its partner pieces. The region and the Fosdick Mountains seemed to have been subjected to shearing in the dextral sense; that is, clockwise, to the right, or present east.

Careful study of deformation patterns in the Fosdick Metamorphic Rocks showed a phase of flattening in the rock, or vertical shortening that occurred as peak metamorphic conditions waned. This we interpreted as due to the rise of the Fosdick metamorphic diapir, a plume of hot, plastic, or soft and pliable rock that worked its way upward through the crust like a stretched-out balloon, pushing and flattening the rocks in its way as it rose along the Balchen Glacier fault (Siddoway et al. 2004).

Along with all these observations, the paleomagnetic data from the area gave evidence that corroborated the strain data and diapir hypothesis. In that study, magnetic directions locked into the rocks around 100

279

million years ago were measured (Luyendyk et al. 1996). The magnetic directions are deflected clockwise to the right and support the dextral shear interpretation of the strain indicators.

As the 2004 study was published, Chris assembled another team to return to the northern Ford Ranges to dig deeper into the Fosdick history. This time, collaborators came from the University of Minnesota, participating in expeditions from 2005 to 2007. Due to challenging weather conditions along the north face of the Fosdicks, they instead studied the south flank with less dramatic outcrops. Despite that obstacle, they found the missing evidence for the core complex hypothesis. They found the detachment fault! Rory McFadden was the lucky student who got to focus his dissertation work on this elusive structure (McFadden et al. 2007, McFadden et al. 2010). Now we understood that tilted magnetic directions in the Chester Mountains to the south of the Fosdicks (Luyendyk et al. 1996) could be explained by both the rise of the Fosdick diapir along the Balchen Glacier fault and tilting of the Chesters above the detachment fault.

The last step was to pull all the data together in a coherent and consistent history (Siddoway et al. 2004, Siddoway 2008). Because the timing we determined for the events preceded the rifting away of New Zealand (or Zealandia), we related the history instead to subduction beneath West Antarctica including Marie Byrd Land and what would later become Zealandia. The diving of an ocean plate under West Antarctica—the Phoenix plate—provided the heat source to get the metamorphism to reach its peak about 130 million years ago. The Balchen Glacier and South Fosdick detachment faults had to be active then so the Fosdick rocks could start to rise up in contact with rocks of the Phillips Mountains. The motion on the Balchen fault included a vertical component to allow Fosdick rocks to move upward and a horizontal component to the right to produce the dextral shear on this fault and the detachment. Subduction slowed and stopped around 105 million years. Geologic studies in New Zealand confirm this (Bradshaw 1989, Kula et al. 2007) and show a period of stretching and extension that started at this time. The Fosdick diapir began to rise during this extension phase, pushing and flattening rocks above and within it while tilting the Chester Mountains out of its way. Southward movement—think of sliding—of the Chester Mountains on the detachment fault created space for the Fosdick core, or diapir, to move upward.

* * *

What about those glacial grooves I observed atop Swarm Peak as told in the story? What do they mean? How and when did they form? Glacial geologist John Stone contacted me in 2000 and suggested work in the Ford Ranges. I directed him to Chris. She invited him to come along with her team on an expedition to collect glacial deposits at different elevations in the various mountain ranges. The goal was to address the question of when and where the peaks of the ranges were last buried under the West Antarctic Ice Sheet. The glacial marks atop a peak indicate thick ice was once there and then retreated. Determining when would provide a data point for the process of deglaciation of West Antarctica (retreat of the ice sheet). Northern Hemisphere ice sheets retreated to their present form between nineteen thousand and six thousand years ago (Yokoyama et al. 2000, Ehlers and Gibbard 2004). Geologists have long suspected that retreat of Antarctic ice sheets have lagged those in the north, but the timing of this is obscure.

Stone performed experiments on rocks previously covered by ice and then exposed as ice receded. The objective was to date the age of exposure of the rock—the surface—and assume this exposure is the time the ice retreated or melted off the peaks. The method measures the amount of an isotope of the element beryllium (^{10}Be). This "nuclide" is formed when oxygen within the minerals of a rock is bombarded by cosmic rays. Ice absorbs cosmic rays and protects underlying rocks. The more of the nuclide in the sample, the longer the rock has been exposed.

The results he, Chris, and colleagues found suggest that the portion of the West Antarctic Ice Sheet in the Ford Ranges thinned by 700 meters over the last 10,000 years. More than half of that thinning took place just 3,300 to 2,400 years ago. This amount of retreat and the timing is much later than what occurred in the Northern Hemisphere. These results point out that West Antarctic ice is still thinning and therefore still contributing to the rise in global sea level (Stone et al. 2003).

In 1995, I published a paper that connected a process—subducted slab capture—to explain crustal rifting in Southern California (Nicholson et al. 1994) with the breakup of the Pacific sector of Gondwana that included MBL, New Zealand, and pieces of submerged continental fragments in the southwest Pacific (Luyendyk 1995). I named Zealandia for New Zealand and the submerged continental pieces, including a vast region of continental crust where the New Zealand islands poked to the surface. New Zealand geologists jumped on my observation. They began a decades-long study of these submerged regions to verify the idea that in fact New Zealand itself was the pinnacle of the submerged continent of Zealandia (Mortimer and Campbell 2014). This culminated in a signature publication in 2017 (Mortimer et al. 2017) that resulted in a global firestorm of media interest. An eighth continent had been discovered and defined. It had been hidden below the waves for tens of millions of years.

Bibliography

Behrendt, J. C., W. E. LeMasurier, A. K. Cooper, F. Tessensohn, A. Trehu, and D. Damaske. 1991. "The West Antarctic Rift System: A review of geophysical investigations." In *Contributions to Antarctic Research II*, edited by D. H. Elliot, 67–112. Washington, D. C.: AGU.

Bradshaw, J.D. 1989. "Cretaceous geotectonic patterns in the New Zealand region." *Tectonics* 8:803–820.

Coney, P.J. 1980. "Cordilleran metamorphic core complexes: An overview." In *Cordilleran Metamorphic Core Complexes*, edited by M.J. Jr. Crittenden, P.J. Coney, and G.H. Davis, 7–31. Geological Society of America Memoir 153.

Davis, G. A., and G. S. Lister. 1988. "Detachment faulting in continental extension: Perspectives from the southwestern U.S. Cordillera." In *Processes in Continental Lithospheric Deformation*, edited by Jr. Sydney P. Clark, B. Clark Burchfiel, and John Suppe, 133-159. Geological Society of America Special Paper 218.

Ehlers, Jürgen, and Philip L. Gibbard, eds. 2004. *Quaternary Glaciations: Extent and Chronology. Part II North America.* Vol. 2, *Developments in Quaternary Sciences.* Elsevier Science.

Korhonen, F.J., M. Brown, M. Grove, C.S. Siddoway, E.F. Baxter, and J.D. Inglis. 2011. "Separating polymetamorphic events in the Fosdick migmatite–granite complex, West Antarctica." *Journal of Metamorphic Geology* 30:165–191.

Korhonen, F.J., S. Saito, M. Brown, and C.S. Siddoway. 2010. "Modeling multiple melt loss events in the evolution of an active continental margin." *Lithos* 116:230–248. doi: 10.1016/j.lithos.2009.09.004.

Kula, J., A. Tulloch, T.L. Spell, and M.L. Wells. 2007. "Two-stage rifting of Zealandia-Australia-Antarctica: evidence from 40Ar/39Ar thermochronometry of the Sisters Shear Zone; Stewart Island, New Zealand." *Geology* 35. doi: 10.1130/G23432A.1.

Luyendyk, B. P. 1995. "Hypothesis for Cretaceous rifting of east Gondwana caused by subducted slab capture." *Geology* 23:373-376.

Luyendyk, B. P. , S. M. Richard, C. H. Smith, and D. L. Kimbrough. 1992. "Geological and geophysical investigations in the northern Ford Ranges, Marie Byrd Land, West Antarctica." In *Recent Progress in Antarctic Earth Science: Proceedings of the 6th Symposium on Antarctic Earth Science, Saitama, Japan, 1991,* edited by Y. Yoshida, K. Kaminuma, and K. Shiraishi, 279–288. Tokyo, Japan: Terra Pub.

Luyendyk, B. P., S. Cisowski, C. H. Smith, S. M. Richard, and D. L. Kimbrough. 1996. "Paleomagnetic study of the northern Ford Ranges, western Marie Byrd Land, West Antarctica: A middle Cretaceous pole, and motion between West and East Antarctica?" *Tectonics* 15:122–141.

Luyendyk, B. P., D. S. Wilson, and C. S. Siddoway. 2003. "Eastern margin of the Ross Sea Rift in western Marie Byrd Land, Antarctica: Crustal structure and tectonic development." *Geochemistry, Geophysics, Geosystems* 4 (10):1090, doi:10.1029/2002GC000462.

McFadden, R., C.S. Siddoway, C. Teyssier, C.M. Fanning, and S.C. Kruckenberg. 2007. "Cretaceous oblique detachment tectonics in the Fosdick Mountains, Marie Byrd Land, Antarctica." In *Antarctica: A Keystone in a Changing World—Online Proceedings of the 10th*

International Symposium on Antarctic Earth Sciences, edited by A.K. Cooper, C. R. Raymond et al., DOI: 10.3133/of2007-1047.srp046.

McFadden, R., C. Teyssier, C. S. Siddoway, D. Whitney, and C. M. Fanning. 2010. "Oblique dilation, melt transfer, and gneiss dome emplacement." *Geology* 38:375–378. doi: 10.1130/G30493.1

Mortimer, Nick, and Hamish Campbell. 2014. *Zealandia: Our Continent Revealed*. Auckland, NZ: Penguin Group (NZ).

Mortimer, Nick, Hamish J. Campbell, Andy J. Tulloch, Peter R. King, Vaughan M. Stagpoole, Ray A. Wood, Mark S. Rattenbury, Rupert Sutherland, Chris J. Adams, Julien Collot, Maria Seton. 2017. "Zealandia: Earth's Hidden Continent." *GSA TODAY* 27 issue 3:27–35. doi: 10.1130/GSATG321A.1.

Nicholson, C., C. C. Sorlien, T. Atwater, J. C. Crowell, and B. P. Luyendyk. 1994. "Microplate capture, rotation of the western Transverse Ranges, and initiation of the San Andreas transform as a low angle fault system." *Geology* 22:491–495.

Richard, S. M., C. H. Smith, D. L. Kimbrough, P. G. Fitzgerald, B. P. Luyendyk, and M. O. McWilliams. 1994. "Cooling history of the northern Ford Ranges, Marie Byrd Land, West Antarctica." *Tectonics* 13 (4):837-857.

Siddoway, C., S. Richard, C. M. Fanning, and B. P. Luyendyk. 2004. "Origin and emplacement mechanisms for a middle Cretaceous gneiss dome, Fosdick Mountains, West Antarctica (Chapter 16)." In *Gneiss domes in orogeny*, edited by D. L. Whitney, C. T. Teyssier, and C. Siddoway, 267–294. Geological Society of America Special Paper 380.

Siddoway, C. S., L. C. I. Sass, and R. Esser. 2005. "Kinematic history of Marie Byrd Land terrane, West Antarctica: Direct evidence from Cretaceous mafic dykes." In *Terrane Processes at the Margin of Gondwana*, edited by A. Vaughan, P. Leat, and J. D. Pankhurst, 417–438. Geological Society of London, Special Publication 246.

Siddoway, C.S. 2008. "Tectonics of the West Antarctic Rift System: New Light on the History and Dynamics of Distributed Intracontinental Extension." In *Antarctica: A Keystone in a Changing World. Proceedings of the 10th International Symposium on Antarctic Earth Sciences*, edited by A. K. Cooper, P. J. Barrett, H. Stagg, B. Storey, E.

Stump, W. Wise, and the 10th ISAES editorial team, 91–114; DOI: 10.3133/of2007-1047.kp09. Washington, D.C.: The National Academies Press.

Stone, John O. , Gregory A. Balco, David E. Sugden, Marc W. Caffee, Louis C. Sass III, Seth G. Cowdery, and Christine Siddoway. 2003. "Holocene Deglaciation of Marie Byrd Land, West Antarctica." *Science* 299:99–102.

Tulloch, A. J., and D. L. Kimbrough. 1989. "The Paparoa metamorphic core complex, New Zealand: Cretaceous extension associated with fragmentation of the Pacific margin of Gondwana." *Tectonics* 8:1217–1234.

Wade, F. A., C. A. Cathey, and J. B. Oldham. 1977. Reconnaissance geologic map of the Guest Peninsula quadrangle, Marie Byrd Land, Antarctica, Map A-7. Reston, VA: U. S. Antarctic Research Program.

Wilbanks, J. R. (1972), "Geology of the Fosdick Mountains, Marie Byrd Land." In *Antarctic Geology and Geology and Geophysics*, edited by R. J. Adie, pp. 277–284, Universitetsforlaget, Oslo, Norway.

Yokoyama, Y., K. Lambeck, P. de Deckker, P. Johnson, and K. Fifield. 2000. "Timing for the maximum of the Last Glacial constrained by lowest sea-level observation." *Nature* 406:713–716.

Acknowledgments

First and foremost, I thank the members of our team who journeyed to the northern Ford Ranges: Dave Kimbrough, now at San Diego State University, Steve Richard, formerly at the Arizona Geological Survey, and my graduate student Christine Smith Siddoway, now a professor at Colorado College. I thank them for their friendship, their dedication to our science objectives, and their commitment to teamwork. Our mountaineers were essential to our success by providing safety oversight of our work. In 1989–1990, they were Alasdair Cain of Kingussie, Scotland, and my friend the late Steve Tucker of Portland, Oregon. In the second season 1990–1991 that appears in chapter 29, our mountaineers were Terry Schmidt of Talkeetna, Alaska, and J.R. Roberts of Mount Cook National Park, New Zealand. Pete Cleary from Christchurch, New Zealand, gave invaluable advice on mountaineering safety that we employed in the field.

Individuals of the US National Science Foundation took a chance on us. We were unproven greenhorns. Herman Zimmerman of NSF is thanked for his demonstrated confidence in our team. Our expeditions had dedicated professional support from many individuals. Those I want to acknowledge are Rick Campbell of ANS, ITT Antarctic Services and ASA, Antarctic Support Associates, along with Ann Peoples and Jill Vereyken with the same contractors. NSF staff in Antarctica helped us put together a program and make it successful. These included Erick Chiang and David Bresnahan. The US Navy aviation squadron VXE-6 provided critical air support for our expeditions. Without them, we would not have reached the Ford Ranges and come back.

Acknowledgments

Staff members of the University of California, Santa Barbara stood by us and helped us along in our often confusing preparations and ongoing efforts. I thank Sara Sorensen, Maureen Evans, Joe Cisneros, the late George Hughes, and the late Bill Bushnell for their unquestioned help.

The team and I benefited a great deal from interactions with geologists who had been working in Marie Byrd Land. They were generous in sharing their experiences, advice, and support. These included Ian Dalziel of the University of Texas, Austin, Anne Grunow of Ohio State University, and John Bradshaw of the University of Canterbury, New Zealand, among many others.

The transition from academic authorship to creative nonfiction was not without pitfalls that required rewiring my psyche and pen. I thank my teachers Shelly Lowenkopf and Toni Lopopolo for considerable help and attention. Shelly gave me encouragement that I could write a book. Toni showed me what lines not to cross. In my learning process, students in my classes at Santa Barbara City College Continuing Education showed me steadfast support and consideration.

My manuscript benefited from reviews by Philip Rosmarin, Ken C. Macdonald, Charles Franck, Matt Leibel, Telly Davidson, Helga Zeigler, Jaye Up de Graff, Connie Hood, and Kathy Oxborrow—thank you!

Thanks go to my editors, Elizabeth Lyon and Toni Lopopolo, and to my agent, Rita Rosenkranz. Alex Novak of Post Hill Press recognized the value of my story. Editorial assistance by Heather King and Madeline Sturgeon at Post Hill is greatly appreciated.

In a multiyear effort such as this book, family members were an essential part of my support system. They made my book possible by never allowing me to give up. I thank my son, Loren, also an Antarctic veteran, for his demonstrated pride in my work, my brother-in-law Philip Rosmarin, for cheering me on, and my dear wife, Susan, who never wavered in her belief in my story or me.

—*Bruce Luyendyk*, Santa Barbara, California, May 2023

Index

Acey-Deucey Club, 242
Aerograms, 147
Airdrop, 85–86, 88, 92, 109, 161,
　　164, 223, 227
Ali, Muhammad, 239
Allen, Woody, 200
Alvin, 144
Amundsen, Roald, 46, 59, 67–68, 99,
　　142, 151, 168, 177–179, 185,
　　207
An Empire of Ice, xvi
Anisotropy of Magnetic Susceptibil-
　　ity, 192, 269, 279
　　AMS, 191–192, 194, 266, 269, 279
Annie, 76, 80, 102, 141, 145, 167,
　　171, 231, 240–241, 244, 253
Antarctic Peninsula, 5, 55, 245, 269
Antarctic Services, 28, 269, 286
　　ANS, 28–31, 37–38, 243, 269, 286
Antarctic Treaty System, 269
Anvers Island, 269
Appendicitis, 101–102, 140
Arizona, 32, 113, 263, 286
Arizona Geological Survey, 263, 286
arrest, mountaineering
asthma, xvii, xx, 3, 14–15, 24, 26, 91,
　　115, 123, 153, 158, 236, 238
Auckland, 244, 284
Avon River, 35

Bainbridge, Beryl, 185
Balaclava, 63, 105, 192
Balchen Glacier, 77, 79, 97, 150–151,
　　159, 164–165, 177, 180, 182,
　　186, 190, 195–196, 198–199,
　　213, 227, 233, 254, 266, 273,
　　279–280
Banzai, Buckaroo, 67–68, 188
Barnes Glacier, 62
Barometer, 100, 122, 163, 171, 174,
　　214
Basalt, 116, 189, 273
Beakers, 6, 28–29, 70, 73
Bear of Oakland, 249
Bee Gees, 136
Berg Field Center, 38, 269
Bergshrund, 114
Big Red, 8, 22, 24, 28, 33, 62–63,
　　162, 172–173, 175, 203, 206,
　　210, 230, 232, 243
Biotite, 132, 273
Birchall Peaks, ix, 148–151,
　　153–154, 156, 180
Bird Bluff, x, 98, 194–199, 201, 203,
　　209, 213
Block Bay, 77, 98, 177, 180, 182, 190,
　　228
blue ice, 74, 177–178, 183, 199, 216,
　　266

Index

Bowie, David, ix

Bradshaw, John, 257–258, 280, 282, 287

brake lever beam, 153

brakeman, 182, 213

British Antarctic expedition, 60

British Antarctic Survey, 235
BAS, 188, 237

Brunton compass, 111, 158, 166

Building 155, 25, 27–28, 234, 240

Bunny Boots, 3, 5, 8, 18, 24, 26, 63, 70

Business Insider, 260–262

Byrd Coast Granite, 119–120, 272–273

Byrd First Antarctic Expedition, 78, 79, 130

Byrd Second Antarctic Expedition, 249,

Byrd Station, 39

Byrd Surface Camp, 37, 82, 100, 164–165, 186–187

Byrd, Admiral Richard E., xviii, 5, 50, 79, 87, 107, 130, 175, 194, 247–248, 250–251

Cache, 77, 213, 217, 220

Cadbury chocolate, 165

California, 21, 53, 55, 76, 103, 246–247, 258, 270–271, 282, 287

Campbell Plateau, xvii, 256, 260–261

Campbell, Hamish, 260–261, 282, 284

Campbell, Joseph, 145

Campbell, Rick, 286

Canterbury University, 257, 287

Cape Evans, 44–45, 47, 49, 51, 56, 59, 62, 234

Cape Royds, 44, 46, 56–57, 62

carabiners, 70

cargo pallets, 221, 230, 254

Carhartt, 20, 28

Challenger, HMS, xvi

Challenger Plateau, 256

Cherry-Garrard, Apsley, 232

Chester Mountains, ix, 93–94, 97–98, 103, 107, 109–111, 113–115, 118, 122, 150, 272–273, 280–281

Chester snowfield, 84–85, 88, 92, 110, 179–180, 217, 229

Christchurch, xxiii, 1, 6–9, 11, 23, 27, 30, 33, 35, 74, 143, 215, 224, 243, 270–271, 286

Christmas, 46, 137, 140, 142, 156, 161–164, 171, 179, 222, 246, 264, 266

cinder cone, 51, 189, 273

Cleary, Pete, 286

climbing harness, 70, 255

Clothing Distribution Center, 7, 270
CDC, 7, 9, 23, 270

Club Med, 25, 232

cold air challenge, 13

Colorado College, 263, 286

combat zone, 82

Condition One, 22–23

Condition Three, 22, 174

Condition Two, 22, 174

constipation, 104

cooling ages, 156, 279

Coordinated Universal Time, 268

cordierite, 155, 273, 276, 278

crampons, 28, 70, 73–74, 114–115, 119, 200

crevasse, xvii, xix, 23, 30, 38, 48, 71, 73–74, 76, 79–80, 87–89, 91, 93, 115, 118, 122, 127, 130–131, 151, 153, 177, 204–206, 208–209, 211, 213, 215, 220, 224–225, 237, 239, 252, 255, 263

Crevasse Valley Glacier, 115–116, 122, 224

Cyclone, 171

Darwin, Charles, xv

Darwinism, xv

David, Edgeworth, xvii, 61

Deep Field, xvi, xx, 34, 37, 69, 152

Delta, 69, 76, 165–166, 228

Denton, Andrew, 154

Denver, xx, 78

Derelict Junction, 27

dike, 116, 273

Dinty Moore, 140, 222

Discovery Channel, 235

Discovery hut, 51, 59

Dodson, Robert, 263

dog tags, 7, 9, 87, 116

Dostoyevsky, Fyodor, 13

Drake Passage, 55

Dry Valleys, 23, 33, 36–37

duct tape, 64, 128, 154–155

earplugs, 2, 64, 86, 92, 99, 153, 169, 175, 230

East Antarctica, xxii–xxiii, 147, 265, 267, 278, 283

Elephant Island, 55, 59

Emergency Cold Weather clothing, 7, 270

  ECW, 7, 9, 20, 29, 32, 57, 170, 172, 242–243, 270

Endurance expedition, 51, 58–59

English, Robert, 76, 142, 249–251

Ensolite, 50, 128, 233

Enya, 225

Eurythmics, 185

Executive Committee Range, 82

Fanning, Mark, 277–278, 283–284

Farthest South, 59

FDX boots, 5

Feldspar, 119, 272–273

Fitzgerald, F. Scott

Fitzgerald, Paul, 1, 156, 277, 284

FORCE, xx, 3, 9, 38, 62, 104, 110, 117, 135, 144, 190, 214, 221, 235, 257, 265, 269–271, 275–277

Ford Granodiorite, 115, 120, 265, 272–273, 278

Ford Ranges, xix, 3, 6, 33, 37–39, 73, 77–78, 80, 82–85, 88, 92, 96, 112, 156, 175, 270, 275, 277, 279–281, 283–284, 286

Fosdick Metamorphic Rocks, 113, 132, 265–266, 271–272, 274–277, 279

Fosdick Mountains, 33–34, 65, 79, 84, 88, 93, 95–99, 110, 120, 130, 159, 177, 180, 198, 204, 213, 255, 258, 265, 270–272, 275–277, 279, 283–285

French scientists, 147

Fucking New Guy, 6, 270

  FNG, 6, 28, 270

  Fingee, 6

Galley, The, 22, 25, 27, 29, 67, 89, 143, 167, 174, 234, 238, 242–243

garnet, 155, 273

geochronology, 7

Geographic South Pole, 269

geomagnetism, 7

glacial striations, 158

glare ice, 79, 87, 182, 199, 201, 216

Glossopteris, xv

glycol, 92, 131

gneiss, 158, 272–273, 275, 278, 284

Gondwana, xvi–xvii, xx 2, 7, 33–34, 81, 191, 221, 256, 258–261, 265, 271, 277–278, 282–285

Index

granite, 119–120, 132, 197, 272–273, 283

granodiorite, 115–116, 120, 265, 272–273, 278

Greenpeace, 47, 234–235

Greenwich Observatory, 267, 268,

Griffith Nunataks, 196

GSA Today, 260–261, 284

hand lens magnifier, 154

handlebar sledge, 47, 92, 150, 178, 197

Hansen, James, xxi

Happy Camper School, 69

Hawaii, 241, 244

Herbie, 62

Hercules C-130, 1, 2, 3, 8, 9, 19, 28, 79, 80, 84, 95, 96, 142, 148, 232

Highway One, 28, 240

Honolulu, 243–244

Hooker, Sir Joseph, xv

hourlies, 100, 103, 139, 223–224, 226

Hut Point, 51, 59

hydroponic greenhouse, 25, 264

ice axe, 70–71, 114, 131, 159

Icestock, 242

igneous rocks, 65, 119, 272, 275

Indian Ocean, 5, 245

International Geophysical Year, 269

iolite, 273

Jamesway, 25

JATO, 89

Johnson, Nicolas, 203

jumars, 70, 74, 206, 211

Kerouac, Jack, 220

King Edward, 61

KIT, 50, 87, 95, 102, 152, 184, 222, 224–226, 270

Kiwi, 9, 34, 51, 240

La Jolla, 54

latrine, 104–105

Lectroids, 68–69

Lee, E. Hamilton, 88

Linda, 14

Little America, 247, 250

loadmaster, 82

longitude, xx, 165, 267–268

Loren (Luyendyk), 14–15, 140–141, 235, 264, 287

Mac Center, 40, 84, 95–96, 120, 138–139, 146, 163–166, 174, 187, 205, 224, 227

Mac Town, 24–25, 29–30, 32–33, 40, 48, 51, 69, 143, 174, 234, 240, 242

Mac Weather, 40, 100–101

magnetic deviation, 109, 111

mail drop, 164–165

Marujupu, 98, 113, 130–132, 197

Mawson, Douglas, 61

Mayday, 186, 205

McFadden, Rory, 280, 283–284

McMurdo Sound, 11, 20, 22, 27, 34, 44–45, 56, 64, 235

medevac, 212, 236

metamorphic mineral, 180, 221, 265, 277–278

metamorphic rock, 7, 31, 65, 113, 132, 150, 191, 197, 258, 265–266, 271–272, 274–277, 279

metamorphism, 7, 65, 265, 272, 277–278, 280

mica, 132, 273

Mickleburgh, Edwin, 148

migmatite, 132, 150, 158, 177, 221, 265, 272–273, 275, 278, 283

Miller, Dennis, 161, 252

MOGAS, 108, 270

monazite, 138, 274, 276
morphine, 94
Mortimer, Nick, 260–261, 282, 284
Mosher, Dave, 260, 262
Mount Avers, 180, 184–185, 189
Mount Bitgood, 195
Mount Buckley, xv
Mount Erebus, 20, 24, 44, 57, 70, 190
Mount Getz, 224
Mount Iphigene, 110, 150, 153
Mount Luyendyk, 98, 264, 266
Mount Perkins, 98, 165, 190, 273
Mount Richardson, 196, 225
Movement Control Center, 143, 270
    MCC, 143–144, 239, 242, 270

Nansen sledge, 60, 151
Fridtjof Nansen, 27
Nautical Almanac, 111
New York Times, 130
New Zealand, xiv, xvii, xx, xxiii, 1–3,
    7–8, 10, 12, 18, 23, 33–34, 47,
    70, 81, 102, 109, 167, 185, 187,
    221, 240, 243–244, 256–262,
    265, 268, 270, 275–277, 279–
    280, 282–283, 285–287
Nimrod expedition, 59
Non-Commissioned Officer, 39
    NCO, 39–40, 42, 270
Not Physically Qualified, 13, 271
    NPQ, 13–15, 17, 271
NSF Chalet, 20–21, 24, 37, 39, 234

O'Connor Nunataks, 98
OAE, 6–7, 12, 23, 89, 101, 222,
    257, 270
Oates, Titus, 185
Observation Hill, 51, 240
Ochs Glacier, 98, 113, 130, 150,
    177, 179–181, 183, 212–213,
    215–217, 219–220

Officers' Club, 32
    O club, 32–33, 80, 89
Optimus stove, 50, 102

Pacific Ocean, 5
Pacific oceanic plate, 259
paleomagnetism, 33, 138
Palmer Station, 269
paratrooper, 6, 167
Pas-d'-bas, 188
Patagonia, 29
peak 1070, 154, 162, 264
Pegasus, 10
penguins, 58, 61
Peninsula, Antarctic
Phillips Mountains, 77, 79, 97, 113,
    159, 177, 182, 195, 198, 221,
    228, 266, 278–280
Phoenix plate, 259, 265, 280
pitch, 210, 229
plate tectonics, xix, 81, 259–260
Point Mugu Naval Air Station, 84
Point of No Return, 1, 3, 5, 7, 9, 11,
    270
    PNR, 9–11, 270
Polar Grid system, 268
Polar Plateau, 62
police whistle, 40, 116, 123, 210
Priestley, Raymond, 61, 136
Prime Meridian, xxii, 267–268
Principal Investigator, 6, 32, 270
Prusik knot, 150
Pull-Out, 30, 109, 138, 148–149,
    220–225, 227, 229, 231, 233,
    243, 252
Put-In, 6, 30, 32, 34, 37–39,
    69, 84–85, 88–91, 93, 95, 97,
    100, 103, 107–108, 113–114,
    129, 138–139, 142–143, 153,
    173, 226
Pynchon, Thomas, 62
Pyne, Steven, xix

Index

quartz, 119, 272–273
Queen Alexandra, 61
quinzhee, 72
Quonset huts, 25, 33

Rainbow Warrior, 234
Raro, 102–104, 114, 170
Razorback Island, 62
Recce, 38, 73, 78–85, 87–88, 196
Red Cross, 94
Reece Pass Glacier, 98, 196, 203
rock box, 38, 111, 142, 154, 221, 227
rock hammers, 49, 158, 218
Rocky Mountains, 246
rope brakes, 65, 178, 183
Rosetta Stone, 120
Ross Ice shelf, xv, xxii–xxiii, 44, 50, 80
Ross Island, xv, xxii–xxiii, 20, 44, 56
Royal Society Range, 35, 45
rucksacks, 24, 70, 230

San Diego, 53–54, 82, 148, 246, 263, 286
San Diego State University, 286
sastrugi, 79, 84, 88, 91, 93, 114, 118, 143, 146, 152, 230
Science Implementation Plan, 101
SIP, 101
Scott Base, 23, 44, 70, 232, 239–240, 242
Scott Polar tent, 46, 49
Scott, Robert Falcon, xiv, 45
Scott's Hut race, 242
Scripps Institution of Oceanography, 53
Sea Ice Runway, 10, 20–21, 27, 85
Search and Rescue, 9, 40, 115, 205, 271
SAR, 40, 101–102, 212, 271
sedimentary, xiv, 261, 265, 272, 275, 278

Shackleton, Ernest, xv–xvii, 31, 44, 51, 53–61, 136, 207
Sierra Zero-Seven-Zero, 9, 95–96, 100, 121, 138–139, 165–166, 227–228
sillimanite, 155, 273
ski drags, 89, 93, 107, 142, 229
Skido, 47, 188
skua, 25, 234
slot (crevasse), 74, 115, 122, 127, 131, 133, 144, 150–151, 183, 204–207, 209–210, 226, 235
snow petrel, 116, 134, 197, 216
snow pit, 87, 186, 233
solar panels, 163
Sony Walkman, 103
Sorlien, Chris, 264, 284
South Georgia, 55–56
south magnetic pole, 109, 112, 268
South Pole, xvii, xxiii, 18, 23, 42–43, 46, 59, 67, 120, 146, 169, 177, 245, 267
South Pole Station, 224, 269
Southcom, 39, 130
Southern Ocean, xxi, 3, 10, 81, 190, 243, 259
Steger, Will, 245
Stone, John, 281
Strayed, Cheryl, 212
subduction zone, 257, 259
submarine plateau, 33
Sulzberger, Arthur, 77, 98, 130
Survival School, ix, 23, 38, 66–67, 69, 71–73, 75–77, 80, 170, 217
Swanson Formation, 265, 272, 278
Swanson Mountains, 115
Swarm Peak, ii, iv, x, 98, 154–155, 157–159, 161, 266, 276, 281

tectonic deformation, 221
tendonitis, 83, 114, 181
Tennyson, Alfred, Lord, 44–45, 51

Terra Nova expedition, xv–xviii, 44–45, 51, 56, 58–60
The Last Place on Earth, 1, 51, 67
Thompson Ridge, 150
Transantarctic Mountains, xv, 20, 28, 57, 267, 277
Trans-Antarctica expedition, 245
Transition, The, 20, 63, 287
Trivial Pursuit, 136

UB40, 163
UC Santa Barbara, 41, 55
    UCSB, 41, 84, 175, 257–258, 265, 271
University of Texas, 259, 287
uranium, 274, 276
US Antarctic Program, 7, 9, 13, 37, 254, 269
    USAP, 7, 9, 14–15, 17, 23, 100–101, 104, 173, 225, 234–235, 239, 243, 254, 271
US Board on Geographic Names, 264
US Geological Survey, 77, 98
US National Archives, 78, 249
US National Science Foundation, xv, 2, 286
    NSF, xx, 2, 20, 22, 31, 35, 37, 41, 48, 103, 112, 212, 225, 234–235, 239–240, 263, 271, 286

Velcro, 63
Vokey, April, 257

Volkswagen, 200
Vostok, 147
VXE-6, 30–31, 84, 243, 271, 286

WAVE team, 37–38, 190
weather classifications, 22
Weddell Sea, 51, 55
weed whacker, 131
Welles, Orson, 68–69
West Antarctic, xxi, 83, 190, 198, 277, 281–282
    Ice Sheet, xxi, 4, 158–159, 190, 197, 261, 281
    plateau, 83, 221, 261
West Antarctica, xvi, xix, xxii–xxiii, 33, 267, 270–271, 276, 280–281, 283–285
Wheeler, Sara, 117
whiteout, xvii, 10–11, 27, 114, 123, 126–128, 130, 137–140, 176, 179, 190, 230, 252
Williams Field, 10, 232
Wilson, Edward, xv
Winfly, 243
winter-over, xiv, 23
Woods Hole Oceanographic Institution, 144

Zealandia, xviii, xx–xxi, 256–257, 259–262, 265, 271, 277, 280, 282–284
Zim's Crack Crème, 155
zircon, 138, 274, 277–278
Zulu time, 187, 268